Affable Savages

Affable Savages

An anthropologist among the Urubu Indians of Brazil

FRANCIS HUXLEY

Sheffield Publishing Company

Salem, Wisconsin

For information about this book, write or call:
Sheffield Publishing Company
P.O. Box 359
Salem, Wisconsin 53168
(414) 843-2281

1995 Reissued by Sheffield Publishing Company

ISBN 1-879215-27-6

Printed in the United States of America

7 6 5 4 3 2 1

Contents

INTRODUCTION 9
1. THE RIVER GURUPÍ 17
2. THE RIVER AND THE INDIANS 27
3. THE INDIAN POST 36
4. ARRIVAL AT A VILLAGE 44
5. URUBU MANNERS 58
6. CHIEFS 66
7. HUNTING AND EATING 74
8. THE HARD AND THE SOFT 88
9. QUARRELS 105
10. MEASLES AND DEATH 119
11. SEX AND SARACACA 129
12. PUBERTY RITES 141
13. THE MOON'S VOMIT 160
14. GHOSTS AND CHILDREN 174
15. SHAMANS 191
16. RAIN 202
17. MYTH AND THE WORLD 212
18. A CANNIBAL TALE 226
19. THE RITUAL EXECUTION 244
20. LEAVING 265
21. INDIANS AND BRAZILIANS 274
GLOSSARY 283

Illustrations

Map *page* 8

Following page 128

A jungle village
Poling a boat up the rapids
Chico skinning a howler monkey
Making manioc flour
Drinking shibé
Everyday dress
Tero and the author
A woman feeding her tame parrot
Kuata sewing a feather headdress together
Men wearing their best clothes for a feast
Saracaca leading a dance
The name-giving ceremony
A young wife dressed in her best

Kinship chart *page* 287

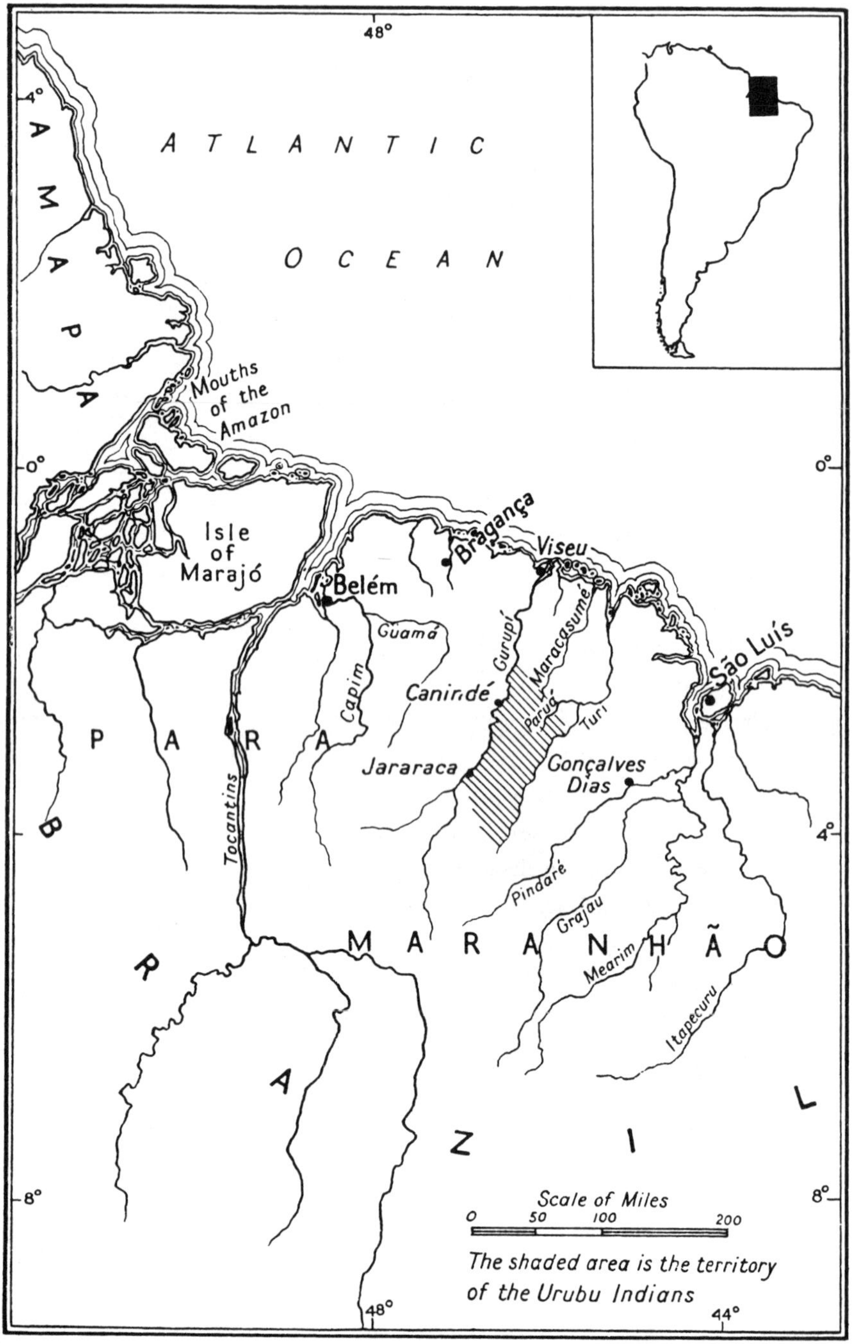
48°
4°
AMAPA
ATLANTIC
OCEAN
Mouths of the Amazon
0°
Isle of Marajó
Belém
Bragança
Viseu
Guamá
Capim
Gurupí
Maracasumé
Caninndé
Paruá
Turi
São Luís
PARA
Jararaca
Gonçalves Dias
Tocantins
Pindaré
Grajau
Mearim
Itapecuru
MARANHÃO
BRAZIL
8°
Scale of Miles
0 50 100 200
The shaded area is the territory of the Urubu Indians
48°
44°

Introduction

Soon after its discovery in 1500, the Portuguese, the French, the Dutch and the English began to settle on the coast of Brazil. They found there a number of Indian tribes, most of which belonged to the same cultural stock and spoke the same language, which is called Tupi; among these tribes the most famous was that of the Tupinamba, of whose customs the early chroniclers wrote in great detail. The Tupinamba, indeed, offered a marvellous opportunity for such study. Whenever a ship arrived from Europe and anchored off the coast, groups of Indians would wait for the new settlers to land in order to propose a novel but agreeable form of alliance: in exchange for trade goods, the settlers would be led off to the Indian villages where they would be treated as honoured guests, being fed, feasted, and given women.

The settlers made good use of this hospitality. Through their hosts they began to understand the new world they had discovered, to recognise the jungle animals and birds, and to adopt the Indians' methods of agriculture and hunting. At the same time, of course, they got to know the Indians, and were especially fascinated by the less pleasant side of their lives: for in the villages were other guests, Indians who had been captured from neighbouring tribes and who were waiting to be ritually killed and eaten as the climax of Tupinamba rites.

The chroniclers are a mine of information, not only on these sensational rites, but on such matters as language, shamanism, marriage, funerals and the organisation of the family; and they also tell of the subsequent history of the Tupi-speaking tribes. As the colonisation of Brazil proceeded, the coastal land where the Indians lived was taken over by the settlers. Many Indians were then killed in war, or were enslaved, or died from the newly-imported diseases of smallpox, measles or the common cold; others escaped into the jungle, or went on enormous expeditions to look for the legendary country of Maran-im, the earthly paradise. (One such party reached Chachapoyas, Peru, in 1549, after nine years of travel up the Amazon; they told of a country they had seen in their travels which abounded in gold and precious stones, a country which the Spaniards immediately took for El Dorado.) The Tupinamba tribe as such, however, disappeared completely, existing now only in the chroniclers' accounts.

These accounts have provided the material for various anthropological studies, chief among which is Alfred Metraux' book, *La Religion des Tupinambas*. I myself did some work on the Tupinamba while I was in England, and became so interested in them that I determined to go to Brazil and do my anthropological field-work among some living Tupi-speaking tribe. Luck was my way: I received a Brazilian Government scholarship and a grant from the D.S.I.R. to cover my expenses; and through Dr. Muller, of the Escola de Sociologia e Política of São Paulo, I was introduced to Darcy Ribeiro, a gifted young anthropologist attached to the Brazilian Indian Service.

The Serviço de Proteção aos Indios is a remarkable institution, whose task is not only to pacify the many tribes of hostile Indians still living in Brazil, but to protect them from civilised exploitation when they have become friendly. It is unfortunate, but not altogether surprising, that this work meets with considerable opposition. The vast jungle of the Brazilian interior has always been a symbol of the country's apparently inexhaustible wealth: a kind of no-man's land that, in spite of its dangers, lures the rubber-tapper, the gatherer of palm-oil, and the prospector to explore its depths. It is not by chance that in

Brazil the verb *explorar*, to explore, has also come to mean to exploit. To this kind of explorer Indians are mainly a nuisance, and since he fears them anyhow he shoots on sight, to kill. The Indians, naturally enough, shoot back, and another jungle war begins.

This attitude towards Indians is one which the Indian Service does its best to reverse. It believes in love rather than bullets, for it refuses to take vengeance on the Indians who kill its workers, and it does its work of pacification by the use of gifts and understanding. An important part of its work, however, is also scientific, and it has done much both to sponsor and to encourage anthropological study.

When I met senhor Ribeiro he was about to go off on a second field trip among a tribe of Indians called Urubu, as part of this anthropological programme. These Indians he told me, spoke a dialect of Tupi; they seemed to have practised ritual cannibalism in the not so distant past—they had only been pacified some twenty-five years ago—and many of their customs and their myths showed great similarities to those of the extinct Tupinamba Indians. When he learnt that this was just the kind of tribe I was hoping to see, he at once invited me to join him in his forthcoming trip. This was an act of great kindness, such as perhaps only another anthropologist can understand properly, for anyone who works with a tribe can hardly avoid feeling somewhat possessive towards it, and thinking of it as 'his' tribe. But it was kind for another reason also. Though I had done some travelling before I had never seen the inside of a tropical jungle, and knew nothing of the practical difficulties of anthropological field-work there. Senhor Ribeiro, however, took all this in his stride, and told me to leave the practical details of the journey to him.

We set off in June 1951, taking advantage of the dry season that ends round about December. From Rio we flew to Belém, where we bought the bulk of our stores;

these senhor Ribeiro arranged to have sent up the river Gurupí to Canindé, the post which the Indian Service maintains there for the Urubus. We ourselves, however, went a different way: flying first to São Luís, on the mouth of the river Pindaré, and thence by boat up the Pindaré to the Indian post of Gonçalves Dias round which another Tupi-speaking tribe, the Guajajaras or Tembé, are settled. From this post we walked through the jungle into the territory of the Urubus, who live between the roughly north-eastern courses of the Pindaré and Gurupí rivers.*

After visiting a few villages together, senhor Ribeiro and I separated; he to continue the work he had begun the year before, I—now with a fairly good idea of how to treat both the Indians and the jungle—to begin my own. When the time came for us to return to Rio, I found that though I had gathered a lot of information I had only the most general idea of what it all amounted to. So, a year later, in the rainy season of 1953, back I went; this time alone, for senhor Ribeiro had other commitments with the Indian Service.

This second trip lasted nearly six months, and since it was then that I began to understand the meaning behind the things the Indians did and talked about, I have used it as a frame for my account of the Urubus. It is of course by no means a complete account, as senhor Ribeiro's forthcoming monograph on these Indians will show. Nor is it really a travel book, though it starts off like one. It is instead an attempt to show how Indians live and enjoy themselves: to show them, in fact, as subjects, not just as the objects of an anthropological study.

To do this I have used as many anecdotes as I could, to let the Indians speak for themselves. Compared to us, Indians have little shame, and the result is a number of passages that are perhaps either improper or horrific. It is difficult to know how to treat such matters, and it may

* See map on page 8.

well seem that Urubu life is basically ignoble, and the Indians are aptly described as savages. Indeed, though this is something of a rude word, it is no use denying that the Urubus are savage. They were well known for their cruelty and vindictiveness in war, in the days before they were pacified; their rites, among which was the killing and eating of an enemy prisoner, were savage with a vengeance; and their manners are often both crude and barbarous. But—leaving aside their saving virtues of hospitality, courage and honesty—this is by no means all that can be said of them. An Indian may well be savage, but this does not mean that he is unprincipled.

Savages in fact have morals, and their world, irrational though it may be, is neither disorderly nor pointless. In such a world nothing can be dismissed as trivial: superstitions, taboos and customs all have their meaning, if one has the patience to search for it. Taboos are especially interesting to the anthropologist, for they arise wherever man sees the supernatural entering—or perhaps emerging from is the better word—the world of nature. This distinction between the natural and the supernatural, between things as they are and as they seem, is somewhat clumsy but still useful. Primitive man is not, on the whole, interested in things as they are: he takes those for granted. It is the image he makes of things that occupies his mind: the image whose form is taken from the material world, but whose purposeful life he supplies out of his own imagination. When such an image has little connection with the real nature of a thing, it is called—by those who do not come under its spell—a superstition: while the taboos attached to it show that the image has a definite mode of action that, as primitive man has so often experienced, can be fatal if not controlled.

This function of magical belief, to act as a kind of umbilical cord between the two worlds of man's experience, is best understood by means of an example. The

Urubus believe that vultures are shamans, or medicine-men, and that the man who shoots them will surely die. The easiest way of finding out why the Indians believe this is to look at one of their myths, which tells of Maír * and the vulture. Maír is the culture-hero of the Urubus, who gave them their customs, inaugurated their rites, and told them how to plant manioc and prepare it as food. Strangely enough, however, it was not Maír but the vulture who originally owned fire; so Maír had to steal it from him, as Prometheus did from Zeus.

The vulture is the original owner of fire in many mythologies. There are obvious reasons for this: the vulture soars high above the world, and is thus the nearest living creature to the sun, that fire of the world; and this 'liking' for the sun is shown by the vulture even when it is perched on a tree, for it will sit motionless for hours, especially at dawn, letting the sun beat upon its half-open wings. More important, it also eats carrion.

Carrion-eaters have always had something of a bad name, a bad name that originally was also a sacred one. At the risk of making it seem that Indians spend all their spare time philosophising, one may compare their belief in the carrion-eating vulture with an early Vedic concept that describes how the world was created by Prajapati, a being who is at once life and death—that is, hunger. 'Whatever he brought forth, he resolved to eat. Because he eats everything, he is called infinite. Therefore he who knows the essence of the infinite becomes the eater of the world; everything becomes food for him' (*Brihadaranyaka Upanishad*, 1. 1. 1). By eating carrion, the vulture becomes a kind of minor Prajapati, since everything in the world sooner or later becomes food for him. It is for this reason that the vulture is said to have been the original owner of fire, for fire is even more notorious for its vor-

* Pronounced Ma-eer. A simplified guide to the pronunciation of Indian words is given on page 283.

acity. But if it destroys, fire also creates: as the sun, it is the source of all life. The vulture shows this two-faced symbolism clearly when he is characterised as a shaman, whose special magic is incarnated in the carrion maggot—itself an image of creation, since it breeds wherever there is decay. Since, however, the vulture never made use of the fire he owned—for the vulture eats his meat uncooked—he is looked upon as a tyrant who refuses to help mankind with his magical powers. Another name for such a tyrant is a sorcerer, who can avenge himself magically on those who shoot at him. It is not surprising, therefore, that the Indians treat vultures with respect.

I have not had the space to discuss this vulture-symbolism in the account that follows, but the meaning behind it crops up everywhere: not only in the figure of Maír, who overcame the vulture and was able to make creative use of the fire he had stolen, but in the whole process to which we give the name of culture. For this process is the way man stops being an animal and becomes human; a process that is described in mythology, though in subjective terms, and which can be seen at work not only in ritual and taboo, the acts by which man sees his own meaning take hold of the world, but also in his undisciplined impulses of savagery, lust or just plain laziness out of which society is eventually formed.

There is perhaps one objection that can be made to this way of dealing with primitive beliefs and customs: that interpretations often look larger than the facts they try to explain, like the genie when it came out of the bottle. One way of meeting this criticism is to compare the Urubus with other Indian tribes. I have had to use comparative material in any case when dealing with ritual cannibalism, since the Urubus were not very forthcoming on the subject: the gap in my information I have filled by summarising the old chroniclers' accounts of Tupinamba cannibalism, to give some idea of what may have happened.

But it is also possible to make more extensive comparisons, for though the Indians of South America may not be descended from Adam, as the first missionaries to Brazil sometimes concluded, there are obvious similarities between their beliefs and European ones. Their mythology, for instance, repeats on a large scale various universal themes, such as the sun's journey through the underworld, or the hero's passage through the clashing rocks or Symplegades; on a smaller scale one can parallel the Urubu myth that tells how the moon and his sister committed incest (page 164) with the story of Cupid and Psyche, and see a faint but interesting resemblance between the way an old man killed Piriangwa in a ritual execution (p. 239) and the myth of Balder, who met his death at the hands of Hother, the blind god. And there are other, more explicit parallels, such as the belief that stone axes are thunderbolts, or that a certain translucent alga is a shooting star—a belief one can find in Dryden, when he writes: 'When I picked up what I supposed was a fallen star, I found I had been cozened with a jelly.'

Such parallels are interesting, though I have not made much use of them, for they show the same genie at work in different parts of the world. But perhaps the only way of being sure that interpretations fit the facts is to try and persuade the genie to go back into his bottle when he has served his turn. The bottle in this case, of course, is the life of the Urubus. This is another reason, if a reason is needed, for describing the Indians by anecdotes; for only thus can one see whether the anthropologist's idea of what an Indian is corresponds at all to the Indian's idea of himself.

CHAPTER ONE

The River Gurupí

THE train left Belém at six in the morning. The engine was wood-fired, and burning embers shot out of the funnel and blew in at all the windows. It stopped to refuel once or twice at stacks of wood cut from the nearby jungle, or at small stations where young boys would invade the coaches: city boys first, selling lottery tickets, soft drinks, chewing gum and newspapers; then peasant boys, with home-made coconut conserves, banana jelly, young coconut milk in bottles, sandwiches and sugary biscuits. On either side of the tracks the jungle grew thickly, cleared here and there for manioc, or sugar-cane, or sisal plantations, with adobe huts under the trees; and sometimes the tracks led straight down the main street of some shabby village, whose occupants would look out of their doors while naked children ran shouting the train along.

At eleven the train stopped for lunch, and I went to sleep. When I woke up at three, we had arrived at Bragança, where a lorry took me and my luggage to the pier on the river. The municipal launch waiting there was a new one, with a powerful new diesel engine; it ran from Bragança, down the river to the sea, along the coastline, into one of the narrow delta mouths of the Gurupí and up to Viseu. We started an hour before sunset, because of the tides, leaving Bragança on its curve of the river, with its mango trees and the palms in front of the large blue municipal building. The journey only became interesting next morning, when we entered the Gurupí: the rising sun then shone with a few bright gleams on the mangrove thickets on either bank, making them look as green as angelica. Behind the mangroves were large stands of swamp trees growing out of the water on stilt roots, and in their branches roosts of scarlet ibises and white egrets.

I recognised Viseu quite easily when I saw it appear on the right bank some time before eight. Darcy Ribeiro and I had passed through it at the end of our trip the year before: we had finally emerged from the jungle at the Indian post of Canindé, taken a boat downstream, and arrived at Viseu in time to catch the launch to Bragança and thence to Belém.

This year, Viseu looked somewhat lusher, for the rains had begun. I disembarked on the wooden pier, had my belongings stored in a small warehouse there, and made my way through the mud to a small pension at the edge of the river. It was small but respectable: it had a tiled roof and a beaten earth floor that was really very clean; out in the garden was the privy and next to it a combined well and bath-hut where you could draw water with a bucket and sluice it over yourself with a small gourd bowl, while the pigs rooted about outside.

The proprietress made me a cup of coffee, that necessary welcome to any guest in Brazil. She was delighted to hear that I was English—the English have been thought much of in South America ever since the days of Admiral Cochrane—and the news that I was about to live with the Indians upriver filled her with a delicious frisson of apprehension. The Urubus have been pacified for some twenty-five years, but tales of their past savagery are not forgotten; besides, the great jungle that stretches at the back of every township in the interior fills people with a certain uneasiness. There is so much still unknown in Brazil, and the unknown breeds fear.

"And how are you going upriver?" she asked me. That was the problem. The officer in charge of the post at Canindé was a lively and intelligent young man called

João Carvalho, who had only just gone upstream in the post's own boat; seeing that I had missed him, I thought I would have to hire a canoe and paddlers to take me from settlement to settlement, like a monkey going from branch to branch, as the local expression had it. Luckily, I was to be spared this expensive and not very reliable way of travelling, for a motor-boat was about to go upriver to a timber camp well above Canindé. It was a dugout canoe fitted with an outboard, and piloted by an exceptionally strong Tembé Indian called Emiliano whom I knew well from my last year's trip, when he had accompanied Ribeiro and myself through the jungle.

I met him by chance when I went for a stroll through the town. He was broke, he was hungry and he was sad; I gave him some money, with the consequence that he got drunk, his favourite relaxation. Luckily he had a companion with whom I could deal—Feliciano, the mechanic, a slender young man with frizzy hair and a dark skin. He was a typical example of the Brazilian peasant, in whose veins runs a mixture of Portuguese, negro and Indian blood; typical, too, in his gaiety and willingness to help. He agreed to take me upriver in a few days; but first, he said, he had some business to attend to further downstream.

I went back to the pension for lunch: fish, caught the night before at the mouth of the estuary, and boiled rice. The odd-job man, who cut firewood and cleaned up around the house, sat on the doorstep while I ate, complaining of life in Viseu. The only connection with the outside world, he said, was by boat, except for an inadequate path through the jungle to Bragança, so that there was little 'movimento' in the town and hardly any amusements.

The few recent matters of interest had all been supernatural. A fisherman had been on the river one night and had suddenly seen Nosso Senhor walking on the water towards him. Our Lord told the fisherman he was going to visit Viseu, next day, in person. When the town knew of this, everyone—except for the small Protestant community who are called 'crentes', believers—put candles in their windows and devoutly waited through the night and then through most of the next day. The crentes laughed at their Catholic neighbours for superstitious fools, and also refused to believe in another prophecy let fall by God himself. An irascible peasant met an old man coming down the path and shot at him; the old man, who turned out to be God, turned the peasant into an enormous hawk and dropped a book by the path. This book was full of prophecies of the coming end of the world; first the hawk-man was going to kill everyone he could find, and then a flood would rise and drown the whole earth.

The crentes took more notice of a fireball, however, since it floated up and down the streets and terrorised the town. A monument was erected as soon as possible by public subscription, in the form of a large cylinder of cement, with some coins stuck over it: this was to placate the fireball so that it should not return. Later a huge cross was set up on the site, the pride and beauty of Viseu, as the local song put it.

The news of the fireball, the hawk-man and the hoped-for coming of Nosso Senhor were all publicly announced by the town's loudspeaker system. Every town worthy of the name in Brazil has one or more loudspeakers set about the main square, broadcasting local news and canned music. Bragança's loudspeaker system went on most of the day; Viseu's only worked for two hours after dark, because the generator was a bit weak, and anyhow used too much fuel. However, what with listening to music during the evenings, and being invited into various houses during the day, to gossip and listen to odd stories, I had plenty to do till Feliciano came back.

He came back one morning in a cloudburst. We were

now well into the wet season, the 'winter', and the main square of Viseu was rapidly turning into mud. But the rain stopped at noon; I then loaded my personal belongings into the boat—most of my stores had to remain behind till Feliciano made another journey, for the boat was not large enough to hold them—and waited till the tide turned. At six we were off.

We stopped two hours later at Nazaré, near the house of a senhor Alfonso. He was the brother-in-law of Mundik (short for Raymundo) Tavares, the owner of the boat; they were in the timber business together, Mundki getting the wood out of the jungle and cutting it at the sawmill upriver, Alfonso putting up the money and looking after the business end. Senhor Alfonso was quite well off: he had a frame house with his own generator for light, and a small general store where the men who worked for him could buy provisions; all round, behind the fringe of jungle on the river bank, was the open *campo* where he raised cattle.

We slept in a small hut by the river. We slung our hammocks between the uprights and, lulled by a pleasing wind that blew in from the river—for the hut had no walls—waited for our supper to cook. It was fresh fish again; and as we ate it, together with handfuls of manioc flour, I felt my journey had really begun. Manioc flour has little to recommend it at first sight: coarse and gravelly, it seems to need a lot of chewing, and its vague nutty flavour hardly seems worth the effort. But it is the great staple in the Brazilian interior, taking the place of bread; and, as one soon comes to find, no meal is complete without it. There is a special way of eating it, which André Thevet, the French cosmographer who visited Brazil in the sixteenth century, observed among the Tupinamba Indians of the coast; 'Ils gardent une estrange façon de faire en mangeant tant ceste farine que autre viande, dautant qu'ils n'approchent jamais la main de la bouche en mangeant, mais la jettent de plus d'un pied loing de la bouche, sans toutefois faillir de prendre la bouché.' Strangely enough, this way of throwing the flour into one's mouth is much more efficient than eating it out of the hand, and one soon learns the knack.

Next morning we went upstream for half a mile, and waited there for some passengers to arrive. Mário, a young peasant boy Feliciano had asked along for the trip, stripped and went swimming in the muddy river. "Look out for the snake!" an old peasant shouted to him from the bank. "It'll come and get you if it hears you!" Here it was called Shiha: further upriver the Indians named it Mai-yu, the yellow snake—yellow because it's thought to be especially active during the rains, when the rivers turn creamy-brown with mud. Mário tried not to take any notice, but he came out soon after, and the old peasant turned to Emiliano and started to talk about, of all strange things, tea. Emiliano showed me his tea: a plant of hashish or maconha which it is strictly forbidden to cultivate in Brazil. That's why they call it tea in front of strangers. "You start smoking it," the peasant said, "and you think, well, it's nothing but smoke, and then after a little you start feeling giddy, you go reeling down the path and the trees turn round and round and round . . ."

"And is that nice?"

"No, sir, it's not! But there are old men here who do nothing but smoke it, they smoke it as if it were tobacco and they don't feel a thing." Emiliano was one such: he bought a little supply here.

Some hours later, our passengers arrived. There were six of them, a family group under the wing of a garrulous middle-aged woman called dona Benedita. She was the sister-in-law of Mundik Tavares, and was going upriver to pay him a visit. Like him, she was plainly of Portuguese stock, sturdy, with a square face and somewhat aggressive manner. With her was Dorico, her thin, pale son who suffered (so she told me) from syphilitic sores on his legs. A sawmill, I thought at first, was no place for him, till I discovered he was going there to fatten himself up a bit, to eat a lot of roast tapir and boiled tortoise-meat. Dona Benedita was already licking her lips in anticipation of the good food the jungle can provide: besides game, there would be fruit to eat, now that the rains had started. She was tired of eating fish.

As pale as Dorico was his sister, Teresinha. Her husband worked at the sawmill, and she had come downriver to have her baby; now she was returning with it in her arms. She was attended by Basilia, a plump, brown young peasant girl, who cooked little messes for the baby and washed its clothes; but she was so overawed by dona Benedita that she hardly spoke a word during the entire journey. Last was Creusa, dona Benedita's eight-year-old grand-daughter, who was bored with the whole adventure.

It was some time before we set off once more: the river was still estuarine, and we waited for the tide to turn. The women fitted their things into the boat, under the 'toldo', the semicircular roof made of plaited withies and palm leaves which kept the sun and the rain off the luggage and off some of the passengers. When the time came to start, Teresinha got well inside, clutching her baby: dona Benedita and Basilia sat on the seat under the edge of the toldo, Dorico and I on a seat further forward. Mário sat on top of the toldo: Feliciano was under it aft, tending his outboard, while Emiliano stood up and steered with his toes on the rudder. Hardly had we set off than the water began to leak in so fast that all our baling couldn't keep up with it: we made hurriedly for shore, beached on the mud, and listened to the old lady start up in a large way—how the water was coming in, and our Lady of Nazaré guard us, and how she wouldn't go an inch further in the boat, not with the water pouring in as it was, and why did sr. Mundik allow such a boat to be sent for her, did he want her to drown?

So we got out, grumpily, unloaded the boat and carried the luggage to the nearest shelter, a small house half a mile away. Its owner, a senhor Jemjim (pronounced jeng-jing) welcomed us and offered us his barn to sleep in; and once we had slung our hammocks there, he called for us to come and have coffee at the back of the house near the kitchen, where some chairs stood round a table on a beaten earth floor. Compared to his neighbours sr. Jemjim was fairly well off. He owned a small store, where he sold rum and tobacco, straw hats and wooden pattens; he grew corn and a lot of cotton, which was now stored in his barn, and he raised cattle. Besides, he had an official position: he was the police delegate's delegate for the district, charged with keeping order during public events.

The last *festa* or holiday had been on the twenty-fifth of December, as he particularised for my benefit; he'd gone to the village where a number of drinking booths had been set up, and told everyone to put away their machetes so that there wouldn't be any fighting. "But, senhor," he said, "it was another Korea! There they were all sitting, with their machetes down their backs, the points well down near their buttocks and the handles sticking out of their collars. But God be thanked, there was no fighting that night . . ."

We had crabs for supper. Beyond sr. Jemjim's garden we saw a torchlight procession going down the path to the river where we had left our boat. These were the people of the village, hurrying to meet the fishermen just back from catching crabs in the swamps below Viseu. Every new moon the crabs come out of their holes and wander over the mud at low tide: they do this on four successive tides, and then are not seen again for another month.

Emiliano straightway went out, bought a large basket of crabs for sixpence, and began to cook them. They were about five inches across the shell: we ate half a dozen each, while Jemjim, his imagination stirred by the thought of the moon having such a power over living things, began to tell me of a man with a similar power of attraction. This was a trader, who operated a boat on the river, buying hides and sun-dried fish from the peasants, and selling them rum, tobacco and provisions. He was a sharp dealer, who always got the best out of a bargain, and the peasants finding themselves cheated would swear they would have no more to do with him. Every time he came, however, they found it impossible to resist him, and they would be cheated again. "He has a magnet in his eye, I think," Jemjim said, "and that's what attracts people. They swear they won't have anything to do with him, but what's the use? They keep right on dealing with him. Magnet in his eye. When he was ill last summer, he went downstream to a friend of mine to get an injection. But my friend said, 'I can't do anything for you, you've too much electricity in your body; an injection would only hurt you'."

Senhor Jemjim, however, was an honest man, and I had no scruples in leaving with him a few more of my belongings, for the boat—even without a leak—was dangerously low in the water. So next morning I stacked them up in a corner of his barn, while Feliciano went round the nearby houses to buy resin and oil from the andiroba tree, to caulk the boat with. He had to pay six cruzeiros a bottle for the oil, which senhor Jemjim thought outrageous; but then, as he remarked, when people need something, others are sure to take advantage of them. That's the way the world is.

We dragged the boat out of the water, turned it upside down, caulked the cracks and the seam between the dugout bottom and the plank sides with sacking, and over that daubed a hot, tarry mixture of resin and oil. The boat, launched again into the river, proved to be quite waterproof; and at six that evening, when the tide came in again, we embarked and continued our journey. The new moon was just going down behind the trees on the northern bank of the river; Emiliano, as a Tembé, should have saluted it respectfully with a cry of 'eh, grandfather!' and asked its blessing. But he was intent on his steering, and it was left to dona Benedita, hunched up beneath the toldo, to ask a blessing on our journey from one of the many saints I could hear her praying to.

CHAPTER TWO

The River and the Indians

THE Brazilian highlands, which form the south-eastern limit of the Amazonian basin, are extremely ancient, and much eroded: the original mountainous plateau has become a rolling upland. The river Gurupí lies at the edge of this formation, cutting through the diminished sandstone layers to expose, here and there, the igneous rock beneath, and eating its way through the small hills, seldom more than sixty feet high, which are all that remain in that region of the once massive highlands. Otherwise the land is flat, and the river flows through it in great meanders. On either side, the jungle presents its anonymous face to the traveller: the trees cloak the bank completely, being hung about with vast threads of lichens, ropes of lianas and small creepers that flower with the advent of the rains—white, mauve or yellow blossoms, that give off the most searching and poignant scents. Above the muddy brown water, which turns into a curious fusty blue where it reflects the sky, kingfishers darted as they were disturbed by the roar of our outboard motor, and the snake-birds—river cormorants—would straighten their serpentine necks to look at us from their perches high above.

It was a monotonous journey, the heat interrupted only by rain, sometimes a gentle shower that was welcome, sometimes a downpour whose swishing veils of water splashed in the river, or guttered lengthily from the trees on the bank: rain warm for the first few minutes, but rapidly chilling one's bones in a slow, trickling discomfort. The squall over, the sun would come out again and in three minutes we would be as hot as ever. There were, of course, respites: at noon we would tie up alongside the bank, clear a place under the trees and cook our lunch; evening usually found us near a small settlement or township, where we would sleep. Otherwise we would camp out, slinging our hammocks between the trees and watching fireflies till we fell asleep, waking up at dawn to find the whole world smoky with mist.

In this way we made our way up the river, willingly entertained by those who lived on its banks: once by a past employee of the Indian Service, once by the two brothers of Mundik Tavares, the owner of the boat, once by the father of João Carvalho, the officer in charge at Canindé. João Carvalho's father was a remarkable man, who had once been a rubber-tapper in the territory of Acre, right in the far western corner of Brazil; but as he approached his sixties, he decided to settle down and open up a patch of jungle for himself on the Gurupí. He had two thousand coffee bushes, orchards of orange trees, mango trees, and banana palms; a plantation of maize to feed his hundred chickens, and some pigs. He fished in the river for catfish and for the notoriously savage piranha, went hunting in the jungle for deer, pig and tapir. It was a good life, as he said, and the only thing he lacked was sugar.

It was a good life, but it had been better; fifty years ago, I was told, a man could take twenty workmen upstream with him, without having to provide any food. At eleven o'clock in the morning, he'd send two men to fish, and back they would come in half an hour with enough for lunch and supper; a huntsman would spend two or three hours in the jungle, and a working party would have to be sent out to collect all the game he had killed. Those were the days at the end of the rubber boom, when even the inferior latex that comes from the Gurupí could be sold for a good price; and during the same period, gold had been discovered in the river, and the place swarmed with panners and prospectors.

But in those days there were also more Indians, some thousands of them—Timbiras (a Gê-speaking tribe) and Tembés on the river, Urubus and others in the jungle. The Timbiras were the first to go, what with fighting against the Brazilians, and disease—I don't suppose there are more than ten alive on the Gurupí today—and then the Tembés began to decline in numbers. Perhaps a hundred of them are left today, but they are fortunately a hardy and peaceful race, who have managed to accommodate themselves to 'civilisation' and yet keep their own souls.

The Tembés are the northern branch of a tribe who call themselves Tenetehara and who have been at peace with the Brazilians since the seventeenth century. Even with the most peaceful of tribes, however, unhappy incidents will occur: one such incident involved the southern branch of the tribe, the Guajajaras, when they attacked Alto Alegre, a town between the rivers Mearím and Grajaú. I heard about it from a lean, grizzled, somewhat cantankerous peasant called Merkid.

"It was when the Capuchin friars lived at Alto Alegre," he said; "if you go there you can see the houses they built, they were solid houses, they're still standing. The Indians decided to attack, several months they were just preparing for the day; they would come in and out of the town all the time, looking about them, seeing how things were. The friars had been told all about it by an old Indian woman—'Look out,' she had said, 'the Indians are getting ready to kill you!' But they didn't take any notice. She told them a second time a month later, then a third time, but they didn't worry about it at all, just a silly story, they said. Then one day when they were all in the church celebrating Mass, the Indians fell on the place and killed every one, all the friars, all the people."

"Everyone? No one escaped?"

"One or two, one or two. But all the friars were dead. And the Indians took away a girl with them, Prepetinha she was called (a diminutive of Perpetua)—a beautiful girl, she came from the handsomest family in Maranhão. Ah, a long time ago, before I was born; my father told me about it. He told me how, twenty years after, Prepetinha was still alive, poor woman; she was with a band of Indians passing near a town, she stripped the bark from a tree and wrote on the tree with charcoal—'Infeliz Prepetinha, ainda viva'—unhappy Prepetinha, still alive."

March 13th, 1901, was the date of the massacre; it was led by a Guajajara called João Caboré, who had been beaten by the Capuchins for refusing to give up one of his two wives. He went round the Guajajara villages showing his scars, and presently found he had fomented a war-party: the Indians were angry with the Capuchins for other reasons. Their children, even babies hardly weaned, would be taken and put in convents, where their parents could never see them; and the Capuchins, as the story shows, were so full of righteous zeal that they could not imagine what wrongs they were guilty of. But sympathy in the towns of Maranhão rests, as is natural, with the Capuchins and Prepetinha; in whose name girls even now will cry out against all Indians that they see.

The Urubus, however, were more savage than either the Guajajaras or the Tembés, and also more difficult to deal with, since their villages were all hidden in the jungle. Prospectors, going upriver in their canoes, would be shot at from the banks; and at night, if they camped ashore, it was more than likely that they would be raided. The prospectors, of course, being no cowards, killed any Indian they saw on sight; and what with raids and counter-raids, the Gurupí soon became one of the more dangerous rivers in Brazil. The Indian Service found that all their attempts at pacifying the Urubus came to nothing, for the Indians had, quite rightly, become suspicious of all strangers; and though the Service continued to put out peace offerings of machetes and cloth, the Indians never accepted them or any other overtures.

The Indian Service at length built two huts, one at Canindé, on a small island in the middle of the river, and another near Jararaca, where the now peaceful Tembés were living. These huts were strongly built, with double walls of split palm-wood, and under the thatch a double ceiling, also of palm-wood; for the Indians would shoot their arrows high into the air, so that they would fall, pierce the thatch and perhaps kill someone underneath.

It was a Tembé Indian called César who told me about all this; a plump, energetic man who spoke in bursts with a slight hiccupy stutter. The hut near Jararaca was built by a man called Dr. Guillerme, who stayed in it with a small group of peasants for several weeks, till things became too hot for him. "They say", said César, "that there were arrows. There were arrows. There were arrows. 'It's no use,' said Dr. Guillerme, 'they don't want to be pacified,' and he went away."

"Did the Urubus fight with the Tembés too, in those days?" I asked.

"Fight? Nothing but fighting. . . . Look, there was a big village of Tembés in those days, somewhere above Jararaca. But, a big village! The caboclos * from the woods came there in a great band, hundreds of them, they killed lots of women and children—the men were all away hunting or fishing—and carried off machetes, axes, clothes, whatever they could find. Off they went into the woods, to pack the things up. Back came the men of the village, kwa, were they angry when they saw everything

* A *caboclo* is properly an Indian-negro half-caste, but is often used derogatorily by townspeople to refer to peasants, while peasants—and 'civilised' Indians like César—use it to refer to tribal Indians.

stolen, the women killed! They went after the caboclos, and then there were bullets. There were bullets! Only two of them got away."

César laughed. "There was one of them lying on the ground with a broken leg. 'No,' he said, 'don't kill me!' The man went up to him and put the mouth of his gun to his arse-hole, right into his tripes, shot the bullet right up!"

"Aie!" I said.

"Kwa," said César, "it was because the Urubus practised all kinds of wickedness. There, when they killed the women and children, they took a stick, sharpened the end, put it into the earth and stuck the children on to the end between their legs, left them like that. . . ." César hung his head and thrust his arms out, like a thief on a cross.

"What?"

"Yes, sir. Stuck them up the arse. And the women too, they stabbed them up the wound of a woman with their arrows, even when they were already dying, stabbed them. . . . Kwa, a caboclo's full of wickedness, why, christians,* they'd kill a christian and cut off his beard, tear it from his face, cut off his prick and everything, spit them all on a stick and leave them there besides the path. Cut his head off too, leave it there looking down the path."

All that was before the pacification. There were so many Indians in those days, he said, it was almost unbelievable, and then of course the measles came and killed hundreds of them. But the pacification itself was all due to an old man, said César—"it was that old Indian who didn't have any hair on his head, only behind his ears . . . an old man. . . . It was him, he was walking about at night, they say he sat down, squatted down on the ground like this, and a tocandira ant bit him, attacked his bag, eh, old caboclo! They say his bag *swelled* up, they say

* Among Brazilian peasants a 'christian' is anyone who lives in a town; the word has lost its religious sense.

he spent all night groaning, then he said, 'I'm going to finish all this, this walking about at night!' In the old days, Canindé wasn't on the bank where it is now, it was on the island. Even there the caboclos used to kill people with arrows: and they walked about at night because they were afraid to show themselves, weren't they at war? 'I'm going to talk to them,' he said, the old man. 'Talk, don't talk, they'll kill you.' 'Kwa, kill me nothing,' he said, 'you'll see.' And it was him who first opened talk with them at the post. Raymundo Caetano it was (a famous Tembé Indian) who was in charge there. 'EH!' the old man shouted from over the river, 'EH!' Caboclos! Off Caetano went with his canoe full of machetes, machetes, machetes, he got to the other side and the old man opened talk with him. 'I want machetes,' he said. Caetano spread machetes over the bank, spread them here, here, here, all over the ground, while the old man went a good way off watching him. Then there were caboclos jumping for the machetes, caboclos like anything! Next day they came again, then it was the affair of hammocks, of knives, of axes, everything. . . . If that old man hadn't been bitten by a tocandira, it would have been a long time till they were tamed. Marajawir his name was, I think."

Marajawir did more than he knew when he accepted the friendship of the Indian Service. In his days, the women —and some men too, if they survived their career as warriors—would live to such an age that their hair would turn white; they would become so decrepit and senile that they would sit all day in the ashes of their fires, just waiting to die. Nowadays you don't see a single Indian who has white hair. Measles epidemics spread through the tribe, with dire results; while 'flu and the common cold have their triumphs too. An Indian gets a cold, becomes feverish, goes for a bathe to cool himself, catches pneumonia and dies.

Those who live on the Gurupí are, naturally enough, quite glad to see the Indians dying out; not because the Indians are troublesome nowadays, but because they occupy a lot of land in which, one never knows, a gold-mine might be found, and which certainly is rich in oil-bearing trees. And even if the Indians are not troublesome any longer, Brazilians still feel uneasy about them and rarely go into their villages, even if they live close by; the sight of an Indian with a bow and a bundle of arrows under his arm is enough to bring all the unhappy past back to mind. Mundik Tavares himself had never been to an Indian village, though he had lived on the Gurupí all his life, and his family had worked for the Indian Service. His great-grandfather, he told me, had helped to pacify the Timbira Indians, and his grandfather had been at Canindé before the Urubus were pacified, and had lived through one of their raids. He fought with them once for two hours, and when at last they went away he saw that his shirt had been torn off his back, and his trousers had come off; his wife was lying dead on one side and his newly married brother on the other. "Those Indians were savages," said Mundik in an unhappy voice, "real savages."

But Mundik has now nothing to fear on the river but the river itself. We picked him up at a small settlement one afternoon, where he'd been buying manioc flour for his men at the sawmill, and again made our way upriver. Others came with him, so that there were fourteen of us in the canoe, which was also loaded up inside with suitcases, baskets of food and cooking things, while outside it was hung about with baskets of coconuts, baskets full of hens, and enormous calabashes. Wet clothes lay on the top of the toldo among seven or eight twelve-foot poles, used when going through the rapids. It is the rapids which are slightly dangerous, and therefore the real fun of a journey upriver; when we came round a bend and saw some ahead, there would be a kind of bustle, and a great sigh from dona Benedita—"O my God," she'd say, "O Jesus Christ, protect us now; Our Lady, give us a good voyage. O St. Sebastian!" Then, suddenly, to Basilia—"Basilia, bale, why don't you bale—look at the water coming in, my God!" Basilia perforce would bale. Meanwhile Emiliano would relinquish the tiller of the outboard and go forward, where he'd take a pole, as would the other men; Feliciano would rev up the motor and aim the boat into the middle of the *pancada* (literally, the blow: where the water rushes white between the rocks) and there'd be shouts, and grunts, and directions, and noises from dona Benedita. If the current was too swift, or the motor packed up, we'd all go over the side and manhandle the boat through the rapids, the water tugging away insistently at our feet as we trod on sharp and slimy rocks; and if the water became too deep, the rope would have to be used. It would be taken up ahead and then we'd have a tug-of-war against the boat which had no one aboard, usually, but Emiliano with a pole—dona Benedita at such moments would be put ashore to try and find a portage. "Embora!" Emiliano would shout, "away!" and then his favourite exclamation, in a high voice—"Pai-té!"

The current was often very swift, and men would lose their footing and be swept away for a short distance downstream; once the boat itself, with dona Benedita aboard, turned broadside to the current and went careering down on to some rocks, and was only saved by Emiliano who brought her up with his pole. "O my God," said dona Benedita, "what fear I have in going up the rapids, with all these men who don't understand and who don't respect anything!" I did not think she was being at all fair; and neither did Mundik, her brother-in-law. "The boats I've lost in the rapids," he said. "It's a difficult river, even nowadays; and I'm getting too old for it."

CHAPTER THREE

The Indian Post

We arrived at the post of Canindé ten days after leaving Viseu. Just below the post, out of sight round the bend, there is an island and some small rapids; then the river comes from the left in a large curve. The post is built on the right, where there is something of a bluff: opposite is a bit of jungle which was felled two years ago but never burnt or cleared for planting, so that the jungle is growing up again. Upstream are some more small rapids, hardly noticeable in the wet season, and beyond them, on the left, the beginning of the path which takes you to the Indian villages.

Canindé is a handsome place. The bank has been cleared of its jungle, and planted with palm trees, with guava trees, and a lone jack tree. The landing place is more convenient in summer than in winter, when it becomes rather steep and muddy; here the dugout canoes are tied up, the peasant men and women of the post bathe, do their washing, and get their drinking water, while the pigs and cattle and dogs wallow nearby. Twenty yards up on the top of the bank is a large barn with a corrugated iron roof and no sides, where manioc flour is prepared; here are large troughs where the previously sodden and decayed roots are pounded up, a press where the poisonous juice is extracted, and the 'oven', a large copper tray some six feet in diameter, supported on a round clay wall, under which a fire is lit to roast the manioc flour and dry it. There are other large pieces of machinery, salvaged from the old Tembé post at Jararaca, no longer in existence; a sugar-cane mill, a rice husker complete with steam-engine, and so on.

A hundred yards away is the main building of the post, a large wooden house with a tiled roof, a wooden plank floor and a spacious verandah. Here João Carvalho lives with his family when in residence; there is a bedroom and an office, an eating room, a large storeroom, and a kitchen annexe, which has a rammed earth floor.

Behind stretches a row of huts thatched and walled with palm-fronds, where the workers of the post—a couple of Urubu Indians, some Tembés and Timbiras, the rest of them Brazilian peasants—live together. In the middle is the earth-closet, a fine building usually occupied by goats.

João Carvalho had just gone down to Viseu again—we had met him on our way upstream—and during his absence the post was run by a dark, stocky and rather silent peasant man called José Pinto. He and his saucy Timbira wife Filomena (Filoca for short) had moved into the post building, and they did the hospitable thing by giving up one of the rooms for me, and brewing some of my coffee for Mundik and the other occupants of the boat. Most of the people round the post came too, to see the new arrivals and hear the gossip. Chic' Ourive came first, a tall, lanky negro with a gift for poking lugubrious fun at himself and the whole world; his real name I have forgotten. Ourive means a gold-panner, referring to his past career on the Gurupí, while Chico is short for Francisco. He had a great knowledge of herbal remedies which he'd got from his mother, famous in her time as a white witch. Only the month before my arrival, he had cured two Indians of snake-bite: both had been bitten by a surucucu, one of the most venomous of Brazilian snakes. Their legs had swollen up till they were three times the normal size (so Chic' Ourive said) but had gone down after the Indians had drunk a decoction of geniparana bark and human urine. At that time he was also looking after a young Indian who was to die some months later, probably of T.B.

Then there was André, a Timbira Indian and the last survivor of his particular group; he was living with a very pleasant woman, dark-skinned and frizzy-haired, called Orminda. Raimunda, her daughter by another man, was looking about for a second husband—her first had got bored and run off some time back. She was being courted by a young man named Shinsh, but wouldn't have anything to do with him, because he had a reputation for being casual. He'd started living with Emiliano's woman when Emiliano was on his journeys and, in a fatal romp one day, had started rocking her as she lay in her hammock: the hammock cord broke and she fell out on to a large iron pot, broke her hip on it and died. Emiliano was both furious and heartbroken: he was saving up to buy a rifle so that he could kill Shinsh, though he was unlikely to do so unless he was very drunk and Shinsh was foolish enough to be around then. Shinsh, I found out later, got his name from a typical Brazilian incident—his real name is Benedito. He'd got embroiled in some quarrel, and didn't dare to appear in public without a machete at his side: being too poor to own a scabbard for it, he went into the jungle, cut down a young palm and took the spathe of the new leaf, called shinsh, and used that instead.

These, and a number of others, came to share our coffee and smoke our tobacco—there is a chronic shortage of both in the interior—and at last to see Mundik off on the last leg of the journey to his lumber camp. The boat was loaded as heavily as it could be with people and baskets of manioc flour that Mundik had arranged to buy here; and newly cut poles were laid on the top of the shelter. Feliciano spent the usual five minutes trying to make the motor come to life, and at last they were off, dona Benedita waving an apprehensive farewell, for the Tapirussu rapids lay ahead of them, the largest and most violent on the river.

Feliciano was to return in a week, bringing with him another Chico, a young Tembé from Jararaca who, I was told, would make me a good interpreter. In the meantime, however, I had enough to do. First, there were a lot of presents to give away. This is an almost prescriptive custom at an Indian post: the small gifts used to sweeten the Indians from time to time cause a lot of jealousy among the peasants, who are often in worse straits than the Indians, being nearly always in debt and hard put to it to find the most necessary of comforts, such as clothes, knives or german-metal plates. For this reason I had brought a lot of trade-goods with me: needles and cotton thread, small mirrors, small knives, cigarette papers, tobacco, fish-hooks and line, soap, which I gave indiscriminately to Indians and peasants; more expensive things, such as cloth, machetes, hammocks and so on I reserved either for my special friends, or as objects of barter. And lastly—perhaps the most popular gift one can make in the interior—I had brought as many kilos of minute porcelain beads, made in Czechoslovakia, as I could afford. These beads are fairly expensive, costing about four pounds the kilo, but are worth every penny.

Giving, however, is luckily not all one-sided. Orminda offered to do my laundry; I was brought some eggs by another woman, Chic' Ourive came with a bunch of bananas, and Engaí brought me some jungle fruit.

Engaí, as José said, is a serious case. He is an Urubu, the brother of a chief; a bit deaf, and speaking in an indistinct murmur, as though ashamed of what he has to say. However, he is nearly always in good spirits and is an energetic worker; moreover he is eminently teasable, which was why Filoca liked him. Engaí had fallen in love with a girl called Cipo-putir, liano blossom, and when I brought out the photographs I had taken on my previous trip he searched through them all without finding her, greatly to his sorrow. Filoca mocked at him, but Engaí got his own back on her. She had picked up a photograph of an Indian man, perfectly naked except for the small piece of string that ties the foreskin in the usual way, and was gazing at it abstractedly: and Engaí, seeing this, snatched the photo out of her hand and began to chaff her in a really very civilised way. He put his head on one side, pursed his lips, stared at the photo and at length put his finger idly on the spot. Then he burst out laughing and pointed accusingly at Filoca, who almost blushed.

Engaí was also something of a glutton. One evening he had four helpings of boiled fish and rice, washed down by a litre of shibé—manioc flour steeped in water—and then, in order to shake his meal down, he began to practise his dancing. Hands clasped in front, elbows well out, he went shuf-shuf-shuffling round the floor, into all the corners, his shirt-tail hanging out and a voracious expression on his face; he was soon all worked up, and when José started to play his guitar he seized Filoca and danced with her on to the verandah.

It was the opening night of Carnival, which is celebrated all over Brazil. At Canindé it wasn't much: no drinks allowed, because it's an Indian post, and the music they make themselves—José with his guitar, singing sambas in a curiously refined and affected voice, quite unlike his usual one; Shinsh with another guitar, trying hopelessly to make some harmony; and Chic' Ourive with a tambourine. As a relief from the sambas there were occasional square dances, the men stamping at the women, the women dancing round the men—a wonderful noise, that gently subsided at two in the morning. At breakfast time, Carnival put on another face, and everyone went about with gourd bowls full of water, playing the game of 'invado'. It's a simple game: the woman gang up on the men and try to souse them, and the men retaliate.

It became slightly more complicated when José found some green and blue dye, which he distributed. He then had to barricade himself into the house against the roving bands of women who were after him. He tried to scare them away by taking off all his clothes and then letting the women burst through the door; but they shrieked happily, closed their eyes and threw the water regardless. Engaí was outside, supplying them out of a barrel, till Raimunda came and starting chasing him with a gourdful of very blue water; all would have been all right had not Shinsh started to chase her, flinging his water slap in her face. She put down her baby, which she was carrying, grabbed a broom, and ran shrieking after Shinsh, beating him on his back till he lost his temper, got her down on the ground and unmercifully retaliated. The poor girl is always being beaten. André, her stepfather, was the last to thrash her for her bad temper: he took a stick and knocked her about the shoulders, and Raimunda just went uuh uuh, like a donkey; and then went to hide herself in the privy for two hours.

Besides Carnival, there was another excuse for making a lot of noise at the post: there was an eclipse of the moon. The earth's shadow started to bite into the moon at about half-past seven; and from then till way past nine o'clock, when the moon was still obscured, everyone honoured the occasion with the traditional salutes; ringing bells, blowing whistles and horns, drumming on the iron roof of the manioc house, pounding the floor-boards, shouting, and letting off musket shots. This, doubtless, is to frighten away the monster that eats the moon. Three Urubus, who had arrived that morning on a visit, didn't seem perturbed, no doubt partly because they saw how much everyone was enjoying themselves: next morning, however, they asked me if the moon would be losing itself again that night, and were relieved to discover it wouldn't.

An eclipse of the moon, of course, isn't half as

frightening as an eclipse of the sun. There was one some ten years ago, of which I heard lots of stories. One story was about a child-naming feast some Indians were celebrating at a village: the men were in all their finest feathers, all painted up, and fairly drunk with manioc beer, when the sun began to be hidden. Then there was a shouting and wailing, with the men picking up their bows and arrows and stamping the ground in fear and anger; and when the eclipse was total they panicked, knocking over the huge earthenware jar which held the beer, tripping up over the hammocks, and colliding with tree-trunks as they tried to run off into the jungle. "If it happens again," one Indian told me, "I think I shall die."

Two days after the eclipse, Feliciano came back from upstream, bringing with him young Chico Mané. Chico was about sixteen or seventeen, with the bashful air many Tembés put on among strangers, when their one preoccupation is not to give away too much of themselves. He was short, stocky—indeed, a little plump—with his straight black hair cut short Brazilian fashion, and his brown eyes set a little aslant; he wore a pair of blue cotton shorts, and a coarsely woven shirt.

I fitted him out the same day with an extremely gay sports shirt (which one of the Urubus at once borrowed from him), a pair of trousers, a good knife, half a pound of tobacco and several packets of cigarette papers. He accepted these gifts gravely, without showing too much pleasure; and his only words were to ask if I could give him another knife, for his father. I liked Chico on sight, so I gave it to him.

Feliciano headed downstream the next morning for Viseu, where he would pick up everything I had had to leave behind. He presented me with a water-melon and two ripe mangoes, and his departure was the moment for another invado skirmish. Feliciano won, with bucketsful of nauseous bilge-water, went off into midstream to protect himself, and then had all the women swimming out for a last insult. José spent an hour or so trying to calm them down—he'd spent most of the morning hiding in the rafters, and had got bored. It was his birthday, too: he was thirty-three, the years of Christ, he said proudly. I opened a large tin of guava jelly to celebrate, and made him a present of some coffee.

Three days later I set out for the Indian villages. Chico had said, "Well, I suppose we'll be leaving soon?" and I said, yes, I suppose so—there didn't seem any particular hurry. I spent a couple of days with José making a huge night-line with forty hooks, to put across the river: but we caught little, for the piranha broke five hooks in half, and chopped the line in three places. Dona Orminda roasted some coffee for me, and got a headache and a cold since she went out afterwards into the rain: Chico and Engaí pounded a lot of rice, and José sawed some wood into boards which I could knock up into a table in the villages.

CHAPTER FOUR

Arrival at a Village

When I left, there were eleven people who came with me. I had decided to take another young Tembé to do the cooking and odd chores: a thin lad who had been nicknamed Picher, or Smell, I don't quite know why. He was very lazy, but he stopped Chico feeling homesick. Engaí came too, to carry luggage; Pompeio, another Tembé, had my tin suitcase as a load; and I invited Filoca and Orminda, since they had never been to an Indian village. They brought their children along, and yet another Chico, Chico Gôgô, to carry their hammocks and odd clothes. Last was Pompeio's wife Maria, a real slattern; she would squat down whenever she had the chance, one leg crooked before her and the other beneath her, a badly eczema'd breast hanging out of her cinder-grey sackcloth dress to tempt the baby that squirmed and wept upon her lap.

We paddled over the river in a large dugout, and apportioned the loads at the other side. Engaí pretended he wouldn't carry anything—"O Engaí," said Orminda, "have you bitten Christ?" But he was just putting on a show for Filoca, and when she'd giggled at him he led off down the path.

The path is smooth and wide for a few hundred yards, till it gets to the *roça*, the plantation of manioc and maize and sugar-cane for the post. The knobbed branches of the manioc plants, with their red stalks and dark green five-fingered leaves, arch over the ground, making it difficult to pass; in addition there are all the prickly weeds that apparently grow only in a roça. Beyond the roça the path becomes a plain track of slightly beaten earth covered with leaves, cleared by machete strokes of any small saplings that try to grow there. We stopped every hour or so, for the women, and lit a cigarette: we sheltered in an assaí palm grove when it started raining, and the boys cut fronds which we used as umbrellas. Beneath the leaves were the hulls of huge rhinoceros beetles, in great profusion: they live in the ground and come out in the winter to breed and die in the open air. Their horns are sometimes mounted in gold, as substitutes for the little hands carved in coral or wood or gold, the thumb between the second and third fingers, that Brazilian women hang down their backs as a charm against the evil eye and other such things.

We hadn't been walking long before Chico Gôgô got bitten on the toe by a jararaca snake. It had been coiled up on a tree-trunk by the side of the path: Engaí had passed it without seeing it, and so had I; but when Chico Gôgô was passing, it had fallen off the log and struck at him. Chico Gôgô was very indignant, as well as in some pain—the snake was luckily only six inches long—"a jararaca's bitten me", he said from time to time, aggrieved and astonished. He advanced on it timidly with a large rotten branch, struck it on the head and killed it. After that he started walking as though nothing much had happened, refusing an injection of anti-snake venom I had with me.

We saw two other small snakes, and a blue-bellied lizard whose bite is said to be dangerous; and, after trying to shoot some howler monkeys in the tree-tops, we arrived at the sarapousinho in the middle of the afternoon, where we settled for the night. (This place gets its name from the sarapo fish, which is abundant in the small stream running there across the path; and because the place is a resting place, or 'pouso'.) Chico Gôgô who, I gathered, had been complaining increasingly of his snake-bite, still refused my help and went to get some geniparana bark, from which he decocted a liquor; on top of that he drank half a cup of kerosene, and declared himself better though he vomited a little blood. He was all right next morning.

This was the second time Chico Gôgô had been bitten by a snake, but jararacas weren't too bad, it was the surucucu that was terrifying. It would coil itself up into a kind of tower some two foot high—it's a big snake, the surucucu, some four or five inches thick—and there it waits for whatever may come along, singing *cucucucucucu* to itself. "Ah," said Chico Gôgô, "when you come along and see it like that, coiled up and you with nothing but a bow and arrow in your hand, it's enough to stop you hunting for the rest of your life." He stared at me with his round eyes, moving his little hairy chin from side to side, till I murmured in agreement.

We were ready to move off again next morning at seven; the women had started gossiping at five, and Picher had the coffee ready by half-past six—black and sweet, to be drunk slowly with a mouthful of manioc flour, and followed up by a cigarette. I was surprised how little I recognised the jungle during our walk, since I had walked that very path several times the year before. The sarapousinho, with its clear stream and assaí palms standing beneath the taller jungle trees, I remembered well: then there was the stony side of a small hill up which the path mounted; a large igarapé or creek, now full of water (the last time I had seen it it had been the dry season) which we waded across; a swampy place full of plantains now all in flower, little spikes of red appearing out of the broad green leaves—mboi iji, Engaí called them, the plant like a snake, refusing to touch one lest a snake should bite him. I remembered a small stream with a sandy bed, which the path led on to for twenty yards before it turned off again into the jungle, and a place where the path was covered with small pebbles. Otherwise I didn't remember a thing.

One remembers bits of the jungle for what's happened there. I had no trouble in recognising the tree-trunk on which Chico Gôgô's snake had been coiled when I next went by this path; further on was where Engaí disturbed a nest of hornets—he'd disappeared down the path shouting at the top of his voice, and we immediately scattered into the jungle; the tree with a bees' nest, which we marked for later use; the place where I shot at howler monkeys and missed; the place where I'd killed a tatu rabo de couro, a leather-tailed armadillo—all to no purpose, it turned out, for it was said to dig up dead bodies ("Credo!" said Filoca, looking at its enormous claws) and eat them, and so no one would touch it; the place where Picher stumbled and spilt all the manioc flour over the ground; and the place where there lay stretched the magnificent but headless skeleton of a jaguar, killed some three months ago by an Indian. He'd taken the head with him for the teeth, which would be strung as a collar.

At one o'clock, Engaí stopped at another small stream where we waited for the women to catch up with us. They arrived twenty minutes later; Filoca had hurt herself by falling on to a stump and Orminda had stubbed her toe. Pompeio didn't arrive for another half-hour: he had fallen with my suitcase on top of him and bruised his ribs, and Maria had slipped and fallen on top of her baby; all three, Filoca said, were stumbling and weeping along the path. I was quite bruised myself, from stumbling over roots: it took me several months to learn how to walk properly in the jungle.

From this stream I gradually began to remember the lay of the land. Soon we came to what is called a capoeira, a piece of the jungle which has once been cut down and used as a plantation and then been left for the jungle to overgrow. We came out of the jungle into a fringe of this capoeira, and from there into the roça proper—a strangely neglected roça, I thought to myself, with the grass two or three feet high, on the path. It belonged to Anakã-puku's village: I'd spent a week with him once, and especially remembered his hut because of the cockroaches in it.

When you approach a village, you make a noise. You shout; you beat fallen tree-trunks with a stick; if you have got a cow's horn, you blow it; if you have a gun, you fire it into the air. This is only polite in a country where the only place you can see more than twenty yards is in a village clearing, and where a figure seen in the distance might conceivably be that of an enemy Indian. Besides, the noise gives the villagers time to beat their dogs into submission or tie them up, so that you won't get bitten.

After some minutes we heard answering shouts, but not from the direction of the village. The village, we found, was deserted; the grass had grown over the dusty places between the huts, and the huts, with their walls of split palm, had their doors battened up. Pineapple and sisal plants thrust their leaves up in rows from beneath fallen tree-trunks, and the path to the stream was blocked by a mass of hairy pumpkin plants. We sat down in the shade of a cashew tree and after a little started to knock its fruit down with poles. They're good to eat; sweet, with much juice, and a little acrid.

* * *

A deserted village looks a bit sad, but is a village for all that; a new village has a desolate air of rawness about it, like a slum on a bomb-site. Anakãpuku's villagers had all moved to a site some five minutes away (except for Anakã-puku himself, who was starting a new village further off), and had set up small huts for themselves in a sea of felled tree-trunks. The foliage and the lesser branches had been recently burnt—the ground was still grey with ash—and nothing had been planted. Nor had the jungle yet healed the wound made in it. The pale bony trunks of the trees fencing in the clearing looked strangely naked and unnatural.

Pari—who, hearing the noise we made, came to lead us to the new village—was an old friend of mine. He must have been about forty-five years old; a lean man, starting to get paunchy, with a well-shaped nose, a wide mouth with thin lips, and small but intelligent eyes. The year before he'd come to the village where I had been staying, and had moved in on me; for his wife had just died and, having no other responsibilities, he wanted a holiday. It was now my turn. I entered his hut, took off my rucksack and dumped it on the ground, then sat down on top of it. After a few minutes Pari gestured towards the house-posts and told me to sling my hammock.

We all slung our hammocks in his hut, and crowded Pari and his wife to one end. Chico took up the fire from the middle of the hut and laid it again in a corner—four small logs, arranged star-wise—while Picher took the clay pot down to the stream for water. The only other furniture consisted of three old tortoise carapaces, used either to squat on or as a bowl for the dogs to eat out of, and the end of a large tree-trunk which protruded into the hut. That was useful as a bench, and for cutting tobacco on. A bow and three arrows were thrust into the leaf thatch, to keep them out of the way. Pari's other belongings were hidden inside a cage housing one of his two dogs, a fierce bitch about to have puppies.

The village was a very small one. Only sixteen Indians lived there, in four huts and a couple of rather mean shelters hastily put up, and soon they came to Pari's hut to see us. I remembered the very first village I had ever stayed in: it was in the heart of the jungle, and few of the Indians had seen a white man so close before. It took them two or three days to get over their curiosity, and during that time Ribeiro and I (for I was then with him) were surrounded by a quiet audience, the men lazing in hammocks nearby, the women sitting on the ground with their children, watching our every movement. It was an unsettling experience, and the women made it worse by shielding their infants' eyes with a hand whenever I looked their way, or warning their older children that if they didn't behave I'd come and eat them. The men, however, were more friendly and assured, and would often leave their hammocks to come and sit in mine, where they would exclaim over my pale skin, stroke my hair and wonder at its comparative fineness.

My welcome this time was luckily less embarrassing. Living only a day's march from Canindé, the Indians were all familiar with white men, and one or two of them remembered me from the year before. They greeted us warmly, and Pari—who knew his duty as host—offered us some fruit that lay in a pile on the ground. They were bacuri fruit, some six inches long: their thick husk must be pounded with a stick before it breaks, and inside are a number of large stones, surrounded with a juicy white pap. The taste was delicious; acid, piquant, a bit custardy, with a flavour of lychees and apricots, perhaps; and something that reminded me distantly of grease-paint.

As we ate, I looked about me. Pari sat in his hammock, watching us; he was wearing a shirt and pair of trousers, with his hair cut short, Brazilian style. His wife, Mbeiju, stood just behind him, rather timidly: she was seventeen or eighteen, I guessed, and still gawky, with fat protruding nipples and no breasts, no hips either. She was quite ugly, though in a pleasant kind of way, and when she got used to us she became rather loud-mouthed and—luckily for me—a great teller of gossip.

Then there was Toí, looking at us from his seat on the tree-trunk that came into the hut. He wore trousers, like Pari, but usually only when there were visitors about: clothes to the Indians have become a matter of prestige, and a man will sport a pair of trousers with the most unfortunate holes in them, in order to do you honour. But he had his hair cut Indian fashion, a fringe at the front and the rest hanging almost down to his shoulders. The lobes of his ears were pierced, and from them hung small *nambi putir*, 'ear-flowers', of blue and black feathers elegantly stuck together. There was also a hole through his lower lip, and in it he had poked the end of a yellow feather; during ceremonies he would substitute a great ornament of many feathers, blue, black and red, for this single feather picked up at random.

Toí was a brisk little man, wiry and well built, with excellent manners; friendly, but slightly reserved. Not for him, as I found out later, the usual plaintive begging to be given beads, or knives, that so many Indians practised: he took gifts as gifts and so kept his self-respect. The women in his family circle, however, cared little for such masculine proprieties, and often came round wheedling me to give them an extra something. Terepik, his wife, being still quite young, was not as forward as her mother, Paijé; and she was also a little shy, keeping well behind her husband on his tree-trunk seat. She had a skirt of blue cotton, the waist-band coming very low down on her hips, outlining the lower curve of her full belly, the material caught up into a number of pleats in front; ear-pendants like her husband's hung from her ears, and there were faded stripes of black paint over her arms and body. Her mother, a stout, leathery woman, was not at all abashed by my presence, but came right into the hut and leant against one of the house-posts; in honour of my presence she had put on a shabby cotton frock that some Brazilian woman must have given her.

"I've never eaten bacuri fruit before," I told Pari conversationally. Pari was amazed, and so was Toí.

"Are there no bacuri trees in your country?" Toí asked, and when I said no, he turned to his wife—"There are no bacuri trees where Francik lives!"

I laughed slightly at the way he pronounced my name: all Indian words are accented on the last syllable and never end with an *s*, so he substituted a *k* to make my name easier on his tongue. "*Francis!*" I said, but though he repeated my name several times he stuck to the *k* at the end of it.

"Are there peccary in your country?" he went on. "No," I said again, and went on saying no for quite a long time—no peccary, no tapir, no jaguar, no paca or agouti (two species of large, rabbit-like animals, good to eat), no macaws, no parrots, no bananas, no sweet potatoes no surubim fish or piranha.

"Mai!" said Toí, his lively face going blank with astonishment. "Then what do you eat?"

So I told them; but Paijé, idly rubbing her back against the house-post, thought such food could hardly be nourishing. I was tall, she remarked, but I was also thin —which was true enough, when I compared her large and admirable belly with mine. "I'll bring you a shibé every day," she said; "soon you'll get fat, when you go home your mother will say, 'You're fat!' " She sent one of her young grand-daughters, who had been hiding behind her, to make a large shibé for us all, and presently it came in a big earthenware pot, a gourd bowl floating on top. Shibé is made from manioc flour: the flour is washed till all the fibres float off, and the starch comes out; then it is left in water to steep till soft. The gourd is dipped into the pot, the mixture stirred round and then scooped out; you put the gourd bowl to your mouth and, using a finger to keep the heavy flour from settling to the bottom, you gulp it down. I've seen some Indians put away a litre of shibé in ten seconds or so; I would take my time.

I was handed the gourd bowl full of shibé, and drank it down slowly. "Saé puku's drinking shibé!" Tero cried out with satisfaction. Tero was Toí's son-in-law, a youth in his late teens, with a fluffy look about him, but a good hunter; a gay, irresponsible fellow, always up to tricks. "Saé puku!" he said, again. Saé puku means long man, a nickname that was always being put on me. He dragged me out of my hammock to measure himself against me, and found that he hardly came up to my shoulder, which caused him a mortified admiration.

"You're my father," he said suddenly.

"All right, child."

"Father, give me some tobacco?"

So I reached for my tobacco pouch, and made cigarettes for everyone who wanted one. Tero and I remained on this familiar footing throughout my stay in the village, though he would sometimes elevate me to the rank of grandfather, as a mark of added respect, or sometimes just to tease me. The others sometimes called me grandfather as well, especially when talking about me to their children.

I returned to my interrupted shibé, and finished it. It is a most satisfactory drink. Nothing's better when you're hot, hungry and thirsty: a well-mixed shibé, with its nutty, slightly acid, perhaps slightly fetid taste, does as much for you as porridge to a Scotsman on a cold morning. The Indians take a shibé whenever they feel a bit hungry—they have one early in the morning, when the sun rises; another at nine or ten; another before lunch; one in the afternoon, another in the evening and a last one if they feel like it before sleeping. It is largely shibé that gives the Indians their rounded, portly bellies—shibé and dry manioc flour that swells a bit inside you when you've swallowed it.

It was Paijé who had handed me my bowl of shibé, so I passed the empty bowl back to her. This was more than good manners: if I hadn't, both of us would have got stomach-aches. In the succeeding days, when Paijé brought shibé in an enormous bowl for all of us, I'd drink what I wanted and would pass it on to Pari; Pari would pass it on to Chico, who'd pass it to Picher, who'd pass it to Tero; then the bowl would be passed back in the reverse order for me to return to Paijé. Once, when I left to go hunting before this circuit had been completed, the bowl lay in our hut all day, even though Paijé had need of it: it was to me that she had given the bowl, and only from me would she accept it again.

The shibé circulated, and the other two men who had huts in the village came to look at us. One was Taperé, Paijé's brother, who lived in a small hut next to Pari's with his wife and small daughter; all three were rather ill, and I rarely saw them. The other was a brash young man called Tahi, whom I distrusted on first sight, for no particular reason; he lived with his two young daughters and his mother, and didn't look after them at all well. His mother, besides, was *kaú*, crazy; mean, morose, and infected with gonorrhoea—"my carapua's hot, hot," she would mutter pressing a hand between her legs. I did what I could for her, but she mistrusted me and would turn her back on my approach.

Taperé hadn't met me before, so I told him my name. Then I had to tell him the names of every member of my family—my mother's and my father's names, which he tried to repeat but couldn't, since they contained l's, unpronounceable by the Indians; my brother's name, and the names of his wife and children.

"And how many children have you got?" Pari asked.

I told him, feeling apologetic, that I had no children; I didn't even have a wife. This shocked him slightly, for no Indian thinks himself a man till he is married. "Panem!" he jeered, though in friendly enough way. To be panem is to fail, to have no luck. And then he returned to the attack: had I fought in a war? how many people had I killed? Was I a chief? Was my father a chief? How many machetes did I have in my house back home, how many beads, clothes, guns? When I came, did I travel by aeroplane? Had Papai-hu, Big Father, sent me to visit them? How long was I going to stay?

I fibbed a little. It was unavoidable. I had to make my position plain and I could do that only by making myself out to be a kind of Indian. Certainly, I said, I was a chief, and Papai-hu had sent me to find out all about them; he wanted to know how they lived, to hear their stories, and —said I, brandishing my camera—to see what they looked like. This explanation of my visit to them was a great success, and I was frequently volunteered pieces of information with the express injunction that Papai-hu should be told all about it. They were slightly put out when I told them, in the interests of truth, that we no longer had a Papai-hu, a king, but a Mamai-hu: it is unheard of among the Urubus for a woman to be a chief. But they soon forgot this uncomfortable fact, and I did not remind them of it. They liked to think of Papai-hu as our equivalent to Maír, their culture-hero; and they attributed to him various miraculous powers, such as the ability to make a gun or a machete just by thinking about it—a fancy I did my best to dispel, in order not to devalue the gifts I had brought with me. I myself was sometimes known as Papai-raír, Father's son, and some Indians were quite disappointed to hear that neither I nor Papai-hu were immortal, as Maír was. But my Papai-hu was always thought of with a great deal of respect, and I was often made to give my anticipatory account of the Coronation. The glittering procession through the town, the warriors lining the streets, the crowns and coronets and robes, and the complicated ritual of the coronation ceremony itself: these were things they could understand without difficulty. Talk about railway trains, sky-scrapers, factories or cinemas only bewildered them.

I was glad when the inquisition at last came to an end, and they began to talk with Orminda and the others who had come with me from Canindé. I had forgotten most of my Tupi, and Chico had as yet little skill in interpreting. The language for him was not difficult: Urubu is a kind of simplified and archaic Tembé, and he already knew it quite well for there was an Urubu village close to his home at Jararaca, which he often visited. However, he thought Urubu an uncouth language, with thick sounds and slurred syllables, and when I made him translate word for word into Portuguese both his tongue and his memory would often fail him, and he would sulk. I didn't blame him: the whole business of interpreting is clumsy, and I disliked it as much as he did, though for different reasons. So did my informers among the Indians who, caught up in the excitement of what they were telling me, would be suddenly cut short while Chico relayed to me the poor fragments that were all he could remember.

But Chico's memory grew with practice, and so did his patience. For the first month we listened to Pari's stories, most of which I had already heard; that was good for me, too, for knowing what was coming I improved my vocabulary and could prompt Chico whenever he left something out. By the end of my six months in the jungle, Chico was quite expert in interpreting, if a little patronising, and I had learnt enough to carry on protracted conversations as long as he was nearby to help me with the occasional word. One Indian, when I went to visit him, was delighted with my progress, as though I had at last grown up—"Last year," he said, "when you first came, you didn't understand anything. Now you speak properly, now you know."

If Anakãpuku (who had been chief of Pari's village before he moved elsewhere) had heard this, he would have said it was all his doing. Once, when neither Chico nor I could understand what he was saying—for his speech was slovenly—he got up from his seat, reached for my fly-buttons, good-humouredly undid them and then spat on me. "Now," he said, buttoning me up again, "now you'll understand what I say. When Darcy was here last year he didn't understand either, so I spat on his *rankuaĩ* for him to understand me." It was an effective piece of magic: Ribeiro certainly spoke Urubu quite well.

We slept soundly in Pari's village that night, and next day I settled myself in, nailing the planks of wood I had brought with me to make a passable table, while Chico made a large shelf below the rafters out of cut saplings, on which we could store our things. Filoca and the others stayed some days, eating, talking, going on visits, gossiping, singing at night and asking riddles—"O que é o que é," they'd ask, "what is it what is it, it passes through water and doesn't get wet?" A shadow. "What is it what is it, has a head of fire and a belly of water?" A paraffin lamp. The Indians thought it all great fun, and began to make up riddles of their own, such as "What walks without legs?" A snake. "What has no bones but stands up?" A plant of manioc. But at last Filoca thought it was time to get back to Canindé, and they set off one day before dawn, with many friendly warnings about the jaguars waiting for them in the jungle. "Jaguars, pooh," Filoca and Orminda said in unison, and walked off down the path. I was glad to be left alone, for it's hard to work amid a lot of noise. Besides, they were eating up all my provisions.

CHAPTER FIVE

Urubu Manners

The Urubu Indians call themselves caápor, or wood-dwellers. It is a name they also use when referring to other Indians, whom they divide into two categories—*caápor-té*, true wood-dwellers or savages (a word that also means wood-dweller), and *caápor-ran*, false ones. True savages are like the Urubu themselves, with feather head-dress, ear-pendants, lip ornament and penis-string. False savages are like the Guajaja Indians, who are nomads and have no settled villages.

When referring to themselves they use two words they have picked up from Brazilians: *cambo*, from the Portuguese *caboclo* (properly an Indian-Brazilian half-caste), and *camara*, from *camarada*, comrade. They dislike being called Urubu, which means vulture, a name first put on them by the Tembés, who call them Uruhu tapií, or vulture barbarians. The name seems to come from the Urubu war-cry, 'Uruhu ne u!' 'Vultures eat you!'

They have been living in their present territory for over a hundred years. They have done a lot of travelling in their time; many of their stories are about the north, about the jungle round the river Capim, and they have one story of a man who lived on the river Tocantins to the west. On these journeys they have intermarried with other tribes, and have picked up a few odd myths and customs not common to the Tupi family, of which they are a branch: for instance, they have two culture-heroes in totally different myth-cycles, as well as a feather headdress that looks as though it came from Guiana.

The Indian Service pacified them in the nineteen-twenties. They have had a fair amount of contact with Brazilians since then, and understand the merits of peace; but, though they have given up their ritual cannibalism, most of their original culture is still intact, and what they no longer practise they can at least remember.

One of the customs which they no longer practise I discovered quite by chance, when an old Indian on a visit to me accepted some food I offered and murmured 'hm hm'. Idly, I said 'hm hm' back to him, and he smiled and shook his head. In the old days, he explained, when you handed a man something, you should say 'ti!'—'hm hm' was the ceremonious reply. A man giving something to his brother said 'ko te shi'; a woman offering her maternal uncle a bowl of shibé would say 'peow', and he would answer 'muhin'; while a young man would address his father with a 'yoí'. These ejaculations were as odd to most of the younger Indians as they were to myself, for they had never heard of them; and for days afterwards Tero would gravely offer me a bacuri fruit, or a stick from the ground, just for the pleasure of hearing the exchange. 'Ti!' he would say; and if I didn't reply he would prompt with an insistent 'hm-hm!'

These set ejaculations, which keep friends and relations at arm's length, have a definite purpose, as I discovered later (Chapter 13), but they are on a par with many other formalities which the Indians delight in. One should never use a person's real name to his face, for example, or he'll feel embarrassed: instead one says 'saé', fellow, or uses his nickname. And if you're talking to someone about a third Indian who may be in earshot, you speak of him as the brother of so-and-so, or so-and-so's father; or you can just point in his direction with a pouting thrust of the lower lip and say 'that one there'.

An Indian may have at least three names: his real name, given to him at a feast when he's six months old; his nickname, perhaps given to him for something he's done

—Pinuarã, for instance, the wife of a man in a neighbouring village, was nicknamed Saracaca or mud-turtle because as a child she liked eating earth. (She was only cured by the chief urinating into her mouth.) And then when men and women marry and have children, they're called by the name of their youngest child: Antonio-hu, the present chief of Saracaca's village, with whom I became great friends, was usually called Ari-ru, Ari's father, just as his wife, if she were still living, would be called Ari-mae. (Antonio-hu is a nickname given him by a Brazilian, and means the big Antonio; his real name is Wira, tree.)

Indians avoid calling each other by name especially if they are related. They prefer to use the kinship term at such times—'hẽ ru', a daughter will address her father, and he'll reply 'hẽ raír', my child, or 'hẽ taí-me-ẽ', my infant. When he addresses his grandson he'll draw out the last syllable of the kinship term into a slightly ominous and resonant note—'hẽ rainyoooooong!' he'll say. He'll do the same kind of thing when speaking to his nephew, or niece, or son-in-law—whenever, that is, he has to deal with someone who is younger than he is, or of inferior status. This mode of address, which is now going somewhat out of fashion, was one way of teaching children to know their place: in the old days, a grandfather would take a poor view of a rowdy child.

Nowadays children run about the village and shout their heads off, and no one stops them. They make off with their elders' cigars, and smoke them down to the butt before they return them; and in Antonio-hu's village there were some particularly badly behaved children, for they once devoured a monkey someone had shot, instead of dividing it up between everyone in the village, as they should have done. Pari knew what would have happened to him if he'd done such a thing as a child: a good smacking.

But parents don't usually need to smack their children.

They look after them with exemplary patience—Antonio-hu would interrupt whatever he was doing if one of the children in his village asked him for something—carry them about everywhere, feed them at any time, tickle them and lull them lovingly when they were fractious. However, a man once asked me in the middle of the night to shoot my rifle off, to give his child a fright and stop her crying; and I've seen a few children being beaten quite hard, or having their ears boxed. One of them deserved it—"I wish you were dead!" she'd shrieked out to her sister, and her grandmother was on her immediately for saying such a thing.

The Indians were certainly a lot fiercer in the old days when they warred with the Brazilians. Antonio-hu once mocked me, saying that if I could have seen his grandfather and heard him talk in his loud, hoarse, rumbling, bad-tempered voice—which Indians put on when they're trying to be impressive—I'd have been frightened stiff.

"They were hantan in those days," Antonio-hu said with pride—hantan being a word meaning hard, strong, capable of enduring anything.

One of the most interesting uses of this word is to describe the formal conversation called ne'eng hantan, or hard speech, which takes place whenever a visitor arrives at a village. Visitors in the jungle are always suspect at first, since they may be enemies, and many Indian tribes greet visitors by shooting arrows over their heads and yelling war-cries before they lead them to the village and offer them hospitality.

The Urubu use ne'eng hantan. The first principle behind it is that intentions must always be baldly announced. Thus when the visitor arrives he says—"I'm come."

"Are you come to see?" the chief will ask politely.

"I'm come to see."

The chief invites the visitor to sit in a hammock and calls to his wife to prepare a shibé, and then the ne'eng hantan proper begins. In a loud, monotonous, aggressive voice the visitor starts talking—talking about anything that comes to his mind, where he comes from, what he saw on the walk, a tapir trail perhaps or some monkeys in a tree, how so-and-so has beaten his wife—while the chief will come in at the end of every sentence with a loud, an enormously loud, affirmation: 'ya té!' he'll say, or 'ya té gui!'—'a word indeed', meaning that the speaker has been telling the truth. Sometimes instead of interjecting 'ya té!' he'll repeat the last words of the speaker, drawing them out loud and flatly.

When the visitor runs out of gossip, the chief takes the lead and it is the visitor then who must support the conversation with these iterations, which sometimes manage to sound both ferocious and bored. If the conversation is particularly interesting it will go on till a few hours before dawn, and then start up again in the morning.

I have heard such conversations too between men who knew each other well, since they came from neighbouring villages. One would be lying in his hammock, perhaps, smoking a cigar, while the other might be squatting on the ground whittling a piece of wood into an arrow-point. The interminable conversation would go on, recounting triviality upon triviality in short sentences and unnatural voices, for no other reason than that the men like the sense of occasion it gave them.

But semi-formal conversations like those leave out what ne'eng hantan is called for—its hard or difficult words. A stranger talking ne'eng hantan will maltreat words by leaving a syllable out or adding a redundant one, and will often use a strange vocabulary never heard elsewhere, so that he would be unintelligible to Chico who yet could speak the everyday language fluently. This vocabulary however is by no means secret ; it's just formal language used between men trying to impress each other with their hardness. Women understand it but never use it.

Ne'eng hantan is a manner you put on, like a mask: to encourage yourself and intimidate others. The severest intimidation is the *patuc*, where a man seizes his bow and arrows, rattles them ferociously in his hand, stamps the ground with one foot several times and shouts 'hu! hu! hu-hu!' This is an alarming sight, luckily not often seen, for a man may get so drunk with the joy a good patuc gives him that he may do something he hadn't at first intended.

There are other set manners which act as masks—women, for instance, speak in a special whiny high-pitched monotonous voice when they're telling their children to go and do something, or scolding them for a fault; a voice which gets a little higher at every word and then drops fatally at the end of the sentence. A real, sullen deadly witch's voice. The children use the same voice to answer back in.

But the best masks are paint and clothes. Men wear clothes nowadays in order to look like Brazilians and to cut a good figure, especially when paying a visit or at ceremonies, though they go naked, except for the usual piece of string tied round the prepuce, when they're out hunting or working in the manioc plantation. The women used to wear *kwé yu*, short flaps of woven cotton or tucum palm-fibre worn fore and aft, but nowadays these are thought to be shameful and no woman thinks herself decent unless she is wearing a proper skirt.

But clothes, for a man, have no real meaning, for he puts them on only to show that he possesses them. His feather ornaments do something more important: they give to a man the symbolic qualities to which they refer. The feather headdress is the best: this half-halo of yellow, red and black feathers worn horizontally is the emblem of the sun's orb, and turns a man into a kind of sun. This is plain from their mythology, where the culture-hero makes a headdress out of stone and charcoal to look just like a feather one, and gives it to a scape-grace character who puts it on and immediately bursts into flame. The headdress is worn only by men: if a woman puts it on, or even helps to tie it on her husband's head, the yapu birds which are the emblem of the culture-hero and whose yellow tail-feathers are the most important in the headdress, will peck her eyes out.

The feather collar is also only worn by men. Hanging from it is a flute made from a hawk's thigh-bone. This is sounded at feasts when a child is given its name, and when victories are being celebrated. The woman's collar, perhaps the most beautiful of all the ornaments, has no pipe on it, and is made from the feathers of mythologically harmless birds—the toucan, the howi mê'ê or blue thing, and flycatchers. Feathers from howi mê'ê appear in nearly all their ornaments: blue is indeed their favourite colour, partly because it puts them into mythological relation with the sky and the heroic spirits who live there.

The headdress and collars are all ceremonial wear. For every day, the Indians wear their small blue ear-pendants, wristlets of red macaw breast-feathers—which go with wristlets of black seeds—and, very often, pieces of cotton thread which they tie round their legs, below the knee, to 'make them walk well'.

But the finest mask is paint: black paint to protect you, red paint to vivify. The black paint comes from the large fruit of the genipa tree. You put the fruit into the embers of a fire for a few minutes, to set the juice free : then you cut the top off, squeeze the sides, and dip a makeshift brush, of cotton-wool tied round a stick, into the juice. Then you paint stripes on yourself: a couple of thin stripes down your cheeks, perhaps, and some circles if you're a chief: two wide stripes from your shoulders, over the chest and ending on each side of the navel: stripes down the arm, stripes round the wrists and round the waist and below the knees. The juice is pale and watery at first: after an hour, when it has dried, it turns a wonderful deep blue-black, and stains the skin so that it won't wash off: if you want to get rid of it you have to wait till the skin itself becomes renewed.

The red paint also comes from a fruit: urucu, it's called, small juicy seeds enclosed in a soft prickled husk, like a young horse-chestnut. You can burst the red seed-bodies with a finger, mash them up and paint your face with the juice: or boil them all up with a little oil till you get a thick red paste, which will keep throughout the year, to supply you when the urucu is no longer in fruit. Urucu is painted on the face only: stripes down the cheek, a stripe over the chin, a stripe at the bridge of the nose: or, during really festive occasions, a mask of red is painted round the eyes and over the cheeks and nose—a rather appalling sight when you see it first, it looks so bloody. But that, presumably, is the idea.

Formal manners, feather ornaments, paint: these not only mark the Urubus as being caápor-té, true Indians, but are the signs of an ideal they are constantly trying to realise in their lives—the ideal of being *hantan*, hard, like their culture-hero Maír.

CHAPTER SIX

Chiefs

INDIAN agriculture is not very efficient. It is of the type known as 'slash and burn', because the only way they have of clearing the ground for a manioc plantation is to cut down the trees and bushes and burn them when they are dry. This is not good for the soil, which in the jungle is poor enough anyway: after two crops of tubers have been lifted in successive years, the ground becomes exhausted and another patch of jungle must be cleared for a new plantation or roça, as the Brazilians call it. The old roça may be planted with special canes, which provide the Indians with shafts for their arrows, but in a few years it has disappeared under the encroaching jungle.

When the most fertile parts of the surrounding jungle have all been exhausted, the Indians move away to another locality, build a new village, clear a new roça, and start all over again. The huts in the old village are burnt to the ground; the huge earthenware plate in which the manioc flour is roasted, and the enormous pots used for brewing cauin,* are broken into pieces, and new ones made for the new villages. Pari and the others in his village, however, had built their new huts only half a mile away from the old village: the patch of jungle they cleared had been virgin, and they wanted to live close to the new roça.

When a village is about to move, some of its inhabitants will often decide against accompanying their fellows to the new site, but will go to some other village where they have friends or relations, and settle down there; or they will start a village on their own. Villages are formed first

* Manioc 'beer'. See page 109.

around family ties, and then around the figure of the chief: the more successful the chief is in making his people happy, the more people will come to stay with him. Anakãpuku had obviously not been wholly successful as a chief for, though he was making a new village several hours' march away, none of his old villagers wanted to go with him. Instead they grouped themselves about Toí, the man at the centre of the village relationships. He was the cousin and son-in-law of Pihun ambir, le feu Pihun: Pihun had been the village chief before Anakãpuku, and Toí, when Pihun died, had taken over his family responsibilities. Tero, Pihun's great-nephew, would be the heir to this position of Toí's, and was bringing up one of Toí's daughters to be his wife; his other fiancée was Mahomé, Pari's step-daughter. Pari, a stranger to the village, made things even by marrying Mbeiju, Toí's sister-in-law—a woman for a woman is an ancient axiom among Indians.

I was surprised that no one wanted to go with Anakãpuku, for he was a hard worker, and famous for his courage. It seems that he had quarrelled with several people, Toí among them, for reasons I could not discover; and Toí took much pleasure in quietly slandering him. So did Paijé: she called him a wicked old man, and told me that he'd recently threatened to cut her hands off and toast them in the fire. This was because Anakãpuku's pop-eyed and somewhat gaga son, Shiã, had had an affair first with Paijé, and then with Tahi's mother, who had the pox: so Shiã caught the pox too, which infuriated his father.

All might have been well had Anakãpuku been closely related to his villagers, and shared some family obligations with them. But he had no relatives in the village, and he was apparently not generous enough to overcome the sense of difference this created. It is the business of a chief to be generous, and to give away what is asked of him: in some Indian tribes you can always tell the chief because he has the fewest possessions and wears the shabbiest ornaments. Everything else he has had to give away. Fortunately for him, the Urubu chief is not expected to be so open-handed. He should, however, look after his people, help them to get machetes and knives, see that the roça is large enough to give manioc all the year round; for if he doesn't his villagers will become dissatisfied and leave him. A chief has no authority outside his own family, and can give no orders—everything is done by agreement, and he has only as much influence as he has prestige. His prestige comes not only from his skill in hunting, his bravery and his cunning, but from his generosity. A mean chief will have no villagers.

Anakãpuku wasn't exactly mean, but he liked owning things. He worked extremely hard, gathering resin in the woods and preparing skins, and went to exchange these at Canindé for machetes, axes, knives and cloth: but when he returned home, there was always someone wanting a machete so badly that he had to have the one Anakãpuku had just brought back. ("When an Indian wants, he wants indeed," a Brazilian said to me once: an Indian hungry for gifts is almost impossible to deny.) He was especially disgusted with the Indians who passed through his village on their way to Canindé, for they were always begging things from him, or stealing them if he wasn't at home. And that was the real reason, Anakãpuku told me, that he was building his new village so far away, in such an isolated spot: he couldn't abide beggars.

Anakãpuku was one of the few real capitalists among the Indians, and would bring me presents of tortoises for the express purpose of being given something in return. Hospitality is traditionally offered freely by the Indians to anyone who arrives, and if the visitor offers a gift in return it's not as payment but as appreciation. Anakãpuku, however, wanted to be paid so that he could accumulate a store of things, and this store represented his offence against traditional morality. Paijé implied this when she inveighed to me against him and the wealth of things he possessed—why, she told me, he had a rope of beads as thick as my arm, his wife had a skirt embroidered all over with more beads, there were lengths of cloth hidden away that would stretch from one end of his hut to the other, and then back again; and he had knives, and machetes, machetes, machetes! Paijé, of course, tried to be a capitalist too, but hadn't the opportunity: and she was outraged because Anakãpuku had refused to give anything to her.

Pari was also a chief, but a chief without a village. He would sometimes lament the days when he was a chief in everything, with his own people: in those days, he said, everyone had worked for him, just because he was a chief, and all he had to do was sit at home and enjoy himself. But then a terrible epidemic of measles had swept through the whole tribe, and wiped his village out. So he went to live with Anakãpuku, because they had been made chiefs at the same time by old Pihun, during a great cauin feast. Pihun had asked all the Indians there who they wanted as chiefs, and everyone said, "Pari and Anakãpuku!" Then Pihun asked the two of them, "Do you want me to put a red cap on your head?" "Put the red cap on my head," they answered. The traditional way to do this is for the old chief and the new to sit astraddle in the same hammock, facing each other, each with a hand on the other's head. The chief-to-be takes two of his arrows and gives them to the old chief, one after the other, by passing them over their outstretched arms; and then the red cap is put on his head. This is how Pari and Anakãpuku were made chiefs: in addition, since they had been made chiefs together and were thus almost brothers, they exchanged their bows and every single arrow they had. After that every man at the feast presented them with two arrows each, and received two in return.

That, said Pari, had been early in the morning, when the cauin had all been drunk. He went hunting in the afternoon, and gathered as many yashi tortoises as he could find, brought them back to his own village, had them cooked for him by another man and when the stew was ready dished it out to everyone—the children especially. "Now I'm your chief!" he said. "Be our chief," they replied, "you're *porang*!" and each in turn came to put his hand on Pari's head. Porang is a fine word: it refers to moral as well as physical beauty, and the villagers when they touch their chief's head both acknowledge this virtue in him and take some for themselves. For the head, when crowned with the feather headdress, is the seat of every heroic virtue, the chief then not so much imitating as becoming that most *porang* and deathless of men, the culture-hero Maír. The chief used to be the only man who habitually wore his feather ornaments every day, except when he went hunting, so that he was literally beautiful: and when he fed the village with food he had himself caught, he showed his moral beauty as well. He mothers his dependents with food and gifts, and fathers them with authority. Every chief is a real, a living Maír.

Pari, however, had very little authority, for he was lazy. When I was living in his hut, he always came to eat my food, not I his, which at once struck me as odd; and though he would provide the manioc flour, he was always having to borrow some from Paijé. The flour he and Mbeiju made was usually full of dust and fibres, and often reddish from having been burnt: Paijé's flour, on the other hand, was wonderful, heavy-grained, without dust, and of a pale yellow colour. So Paijé would mock at him from time to time, and so, I regret to say, would Chico; and Mbeiju certainly was not above teasing him occasionally. Mbeiju, in fact, kept him in good order. She was only a young girl, and much relished the idea of having a man to herself, so that she would often jump on him, and make spirited grabs for his rankuaĩ while tussling in a hammock, frequently urged on by Tero and Chico.

"What were you doing all day in the jungle!" Chico asked Pari one evening, when he came back at nightfall after a day's unsuccessful hunting.

"Hunting," said Pari coldly.

"He suruked me under the bacuri tree!" said Mbeiju, ravenously eating a nhambu bird she had half-cooked in the flames.

"Lie!" Pari said.

"O Pari," we cried, "under the bacuri tree! We'll all get stomach-aches tomorrow."

"It's a lie," he repeated, and Mbeiju at last confessed that she'd wanted to be suruked there, but Pari had said no.

"I didn't suruk her," said Pari. "Some men suruk in the day, and then at night they turn to their women and do it again, and again, and again the next morning, but I'm not like that, I only suruk at certain times. When people are hungry they're just hungry, they don't want to suruk. When they're *hantan*, hard, well, that's all they want to do, but when they're hungry they're soft, they just want to go and drink a shibé."

Chiefs can only make love once or twice every three months—that is, ideally—if they do it more often, they are said to die of it. I heard of one chief who'd died that way, even after he'd been warned. But Pari's abstinence came from another form of panem—he was liable to be impotent. Or so his mother-in-law said. Certainly he had no children of his own, but he explained that differently: it was because when he was a young boy, he'd been sitting on the ground in the chief's hut, and had started hacking at the house-posts idly with a knife. He who does that, begets no children: for a chief's house is like the world itself, and its posts keep the sky from falling.

If it wasn't for his red cap, one couldn't have told that Pari was a chief. The Indians call their chiefs Capitã: and one takes it for granted that the word is a corruption of the Portuguese *capitão*, which is how the Brazilians have always called Indian chiefs. The Indians, however, are indignant if you say it's not their word at all, and they point to the etymology they take from it: akang, a head, and pitang, red. Which is why a chief nowadays wears a red cap, made from manufactured cotton. It was only in my last week that I discovered the old name which had, reasonably, the same general meaning as Captain has; Awa kang-iw is their term for it, the head man.

You could tell Anakãpuku was a chief, however. He wasn't very tall, only about five feet six, but was wonderfully broad and thick-chested, and had enormous muscles on him. He had bright, keen eyes in a weather-beaten face full of bold wrinkles, a small nose and a prize-fighter's jaw. I always admired his hands and feet. His hands were large, stubby, but capable of the finest work; and the lines on their palms were deeply incised but extremely simple. He had, moreover, a simian line, that is, his head and heart lines had joined together into a single line running right across the palm. This primitive characteristic he shared with at least ten other Indians whose hands I noticed casually. As for his feet, their toes were well-splayed out through being continually used, and even the little toe was as large as my thumb: the soles of his feet looked like the face of the moon, the skin enormously thick, and cut, cracked, fissured and gullied. He complained occasionally of the thorns that he trod on, which he would extract by pinching the skin together over the top of the thorn, and cutting it right off, repeating this till the thorn was exposed. "I want shoes!" he'd say. "I'm tired of thorns in my feet," and would borrow mine to shuffle around in till they became too uncomfortable.

But he was always a chief. When he asked for a knife

or a machete or some more tobacco, he would ask for them with the same calm as a Yorkshire business man who knows his rights. And he took care of his people, too: that is, of his new people, Kuata and Koso and Nusir. "Give some more beads to Nusir," he'd tell me, "he's wanting some, his wife hasn't any! He wants a knife too." And he made Koso a feather headdress, which Koso gave to me in exchange for a manufactured hammock, because Koso wasn't much good at that kind of work.

Koso also wore a red cap sometimes, though he wasn't a chief. He was the grandson of a *tushau*, which I at first thought was just a war-chief, but later discovered was also the man who ritually executed prisoners in the days when the Urubus still were cannibals. Tushaus used to wear a cap made from the skin of a maracaja or spotted cat. Just before Koso's grandfather died he had asked each of his sons if they wished to have the cap and be tushau in his place. But since the Indians were no longer cannibals, there was no point in being a tushau, and all his sons told him they would much prefer to be made chiefs. So Koso's grandfather took his maracaja-skin cap with him to his grave, as the last tushau. Koso sometimes thought it would be nice to be a tushau, but there is no one now who can instal him—no one to put the maracaja-skin cap on his head—so he hopes that some day he may become a chief. But people call him tushau, because the name is honourable.

CHAPTER SEVEN

Hunting and Eating

SOME days after my arrival at Pari's, I opened my suitcase and brought out my gifts: strings of beads, small round mirrors, needles and thread, tobacco and cigarette papers, and a few knives. I seldom had much pleasure in giving things to the Indians, for they took a very proprietary attitude with my presents to them, assuming that they were theirs by right; I should, I realised later, have tried to act like a newly-made chief distributing tortoise stew about the village as a mark of his position. That way, at least, I would have fed my vanity: as it was, I felt like a bookie after a bad day at the races. Still, the Indians—though they never say thank-you—went away fairly content, especially after I assured them that later I would have cloth and machetes to give them as well.

Tero took his gifts to his hut, where he hid them, and then went off into the woods to get bacuri fruit. He went off singing, singing, singing—what's got into Tero? we wondered, and yelled to encourage him. Soon his singing turned to a shout, and after a few minutes we heard several other men shouting as well, something about a deer, a deer. "Let's go," I said, and took down my rifle. Pari whistled up his brindled hound, who didn't want to move—just like Taperé's dog, the one called Agouti's Penis, who has to be carried into the jungle whenever Taperé wants to hunt.

We found Tero under a bacuri tree, with two other Indians some distance off, all of them with bows and arrows, and all shouting to their dogs as loud as they could. Pari started in as well—kokokokoko, come here, he cried, but his dog wouldn't budge. Pari dragged it by the scruff of its neck down the path the deer had followed, but it still wouldn't budge, it wouldn't even put its nose to the ground. Pari kicked it in disgust, and after we had gathered some bacuri fruit we made our way back again.

Tero went hunting in earnest. I had a cold coming, and didn't feel very well; Mbeiju was pestering me to give her some more beads, which made me bad-tempered. "Eh grandfather," she said, "more!"

I refused. She sighed and began to mend Pari's clothes, putting a bright red patch next to a dirty blue patch, on a pair of very dirty grey trousers. She gave them to Pari and, as he took off his other pair, which also had a hole in them, threw herself on top of him as he lay in his hammock and started an amorous tussle; a tussle which only stopped when the brindled hound joined in with indiscriminate snaps.

Tero reappeared in the afternoon. He said he'd just arrowed a tapir: he'd seen a mother tapir with her half-grown son, he'd put a takwara, an iron-headed hunting arrow, into the son, and shot a wooden-headed arrow at the mother, which had bounced off without penetrating. Then he'd had to come back, since he had no more arrows. No one wanted to do anything about it. It looked like rain.

"Well," I said, "let's go, get a dog to follow the trail." Tero went to speak with Taperé, who was in the flour house: it starting raining in earnest, and went on for an hour. At three it cleared up enough for us to start: Tero and I, Toí following after, Pari, who'd suddenly decided to come, and Picher who usually only liked sleeping.

Tero crossed the clearing and entered the woods on Toí's side of the village, on the east. Toí uses a small stream there to steep his manioc in during the rains: this year, it was still dry. Tero took off his trousers and, to make himself decent, tied a great strip of leaf round his prepuce; it looked like a bow-tie. Thirty yards on, he pointed to a small tortoise which he had found and hung up by a liana from a tree—it was for me, he said. I left it there, to pick up on our way back.

We came to a small natural clearing, where a tree had recently fallen down and one could see the sky. I do not like the jungle as a rule, for it makes little appeal to my imagination, and it is so full of trivial detail that, by myself, I usually lose my way. All those trees, all those bushes! A small-minded place, I would think to myself, where everything looks so much the same that nothing ever seems to happen. Anything that breaks the monotony of greenness is welcome, and the small clearing, with the fallen tree in its wreckage of branches, was at any rate a landmark for me. I had to admit, however, that the jungle through which Tero led us afterwards was full of exciting things: the bird-of-paradise flower, with its orange spikes zigzagging up from a long green stem; little orchids attached to low boughs, with faintly spotted butterfly-shaped petals; the flowers, magenta, white, red, yellow, that had fallen from a height through the undergrowth and lay in a dim maze of colour on the ground; the new pinky-brown leaves of plantains, and many bromelias, now shooting up into vermilion sprouts. The wild cashew trees were already dropping their little crimson fruit, and so were the massaranduba trees, covering the ground with the rotten husks of their gloriously sweet and milky fruit, on which the tortoises fill themselves up after the rains.

Soon we came to a stream, which we had to ford three times, for it wound about considerably. A jungle stream is a pretty sight when its sandy bed is just covered with water and there are only a few dead leaves and tree-trunks fallen into it; with a tattered ribbon of sky above, into which the assaí palms thrust their heads. We walked along it for a bit, and came out close to the place where Tero had shot the tapir. The dogs picked up the trail and very quickly lost it again; Tero and Pari scouted round, poking the ground to look for tracks—it's astonishing what faint marks an enormous animal like a tapir leaves behind it. They found the trail, which led down to the stream; they searched both banks, found where it came out again and twenty yards further on where it led them back into the stream. Downstream for ten yards, out again, then into thick cover; then down another bend of the stream, where they lost the trail for ten minutes. I handed round my tobacco pouch, and we all had a smoke: I thought the hunt was over. But Pari went off to the left, and Tero ahead, and Toí with his dog to the right. I stayed where I was with Picher for five minutes till we heard Toí's excited shouting: we ran in his direction, hearing dogs barking, hearing a tapir whistle shrilly twice, then more barking and two more whistles as the young tapir broke cover. It saw Pari, turned, saw Toí, heard Picher and me behind it, turned again and got Pari's small takwara in its chest; it rolled down a hillock and fell dead in a dry stream-bed. The mother tapir was still somewhere near, angry at the death of her child: we could hear her breaking bushes.

I went down to look at the body. It was a young tapir, but over half-grown, a good size. Toí motioned to me and Picher to drag the tapir out of the stream-bed, up the small hill and down again to the stream where Tero was preparing a carpet of palm leaves. It was too heavy for us: I called Pari to help, and using ear, snout, arms, legs, tail, anything as a purchase, we pulled the beast up the hill, rolled him down the other side and dragged him to his last green bed, where he lay with his feet sticking in the air and six inches of gut protruding from the hole Tero's arrow had made that morning.

Toí took his machete, squatted down behind the tapir and started cutting the buttock close to the tail, but couldn't get through the skin. Pari took his place and with a takwara made a deep incision.

"Fat!" they all exclaimed, looking at the half inch of fat that lay immediately beneath the skin.

"Who's going to cut him up?" asked Pari, looking at me.

"You," I said simply. He shouldn't have had to, really: the man who kills an animal leaves the cutting up to one of his companions, but I didn't know that then. Pari looked about him for a moment, at Toí and Tero who were already plaiting and knotting palm-leaves together to make *peras*, leaf knapsacks, in which to carry the meat home. So he asked for a machete and started slicing.

"Here?" he called, as he pulled out the hind leg and made Picher pull the other; and cut down the groin.

"There!" said Toí.

Pari cut down one groin, then down the other; he sliced across the belly just under the breast-bone, and again just above the penis; then he joined the cuts together and took out the roughly square belly-piece, leaving the guts plainly showing and easy to take out. Then followed a skilful job of butchery, and half an hour later the carcase lay dismembered on the ground.

"You!" he said, throwing a hind leg down at my feet, "for you to carry!" Picher had to carry heart and liver and neck and diaphragm; Toí, the other hind leg, the spine, the flesh from the back, and half the rib-case; Tero, the hips; and Pari the rest, till he decided against the head and threw it down for me. We packed the meat in the peras, tore strips of bark from an embira tree for shoulder straps, and were ready to start.

"Where's the way back?" asked Tero, pointing behind him. "There? there? there?"

Toí chose a course for him to follow, which to me looked at least ninety degrees out of true, for I'd quite lost my bearings. We made our way back at a comfortable pace, waiting for Picher from time to time—Pari imitating him mockingly, gazing fixedly somewhere among the leaves above him, and lifting his feet up as though to climb into bed at each step. I picked up the tortoise Tero had left for me, and we waited at the edge of the roça for Picher, so that we could all appear together.

"Tero, Tero," said Picher, "don't you want your trousers?"

"Bring them!" Tero and Toí shouted together: and Tero started to put them on, but thought better of it, because he wanted to show off his latest invention, his absurd bow-tie affair.

We hallooed as we came into the clearing, and then walked silently through the village to Pari's hut, where we laid down our loads. The meat was Pari's to distribute, because he'd not only killed the tapir, but had cut it up afterwards: quite often the man who kills game leaves the distribution to the butcher. Tero, who'd first wounded the tapir, only got the tail and the end of the spine. Toí kept the meat he'd been carrying, and took it back to his hut. Pari took the hind leg, gave it to Mbeiju to cook, and hung the rest of the meat below the thatch, out of the way of dogs and wild maracaja cats.

Pari went to bathe. Mbeiju immediately began to tell us about tapir hunts which she remembered, how So-and-so had killed a tapir in the river near Canindé one year, how Anakãpuku had killed another a month ago, and had returned to the village boasting that it was a fat one, though it was really as lean as a board; and then Pari returned, and he began. He told us of the one fat tapir he'd managed to kill in ten years or so; he'd killed it about a mile away from Canindé, and had brought it all in by himself, in four loads. "It was fat, fat!" he cried exultingly.

A woman, too, must be fat to be desirable; not gross, but plump and well built, her rich flesh being the true mark of a fortunate and happy life. But that tapir, how fat it had been! Everyone in Canindé came to ask him for tapir oil, till he hardly had any left for himself.

"Tapir fat," Picher told me, "is what you use when you've got rheumatism. Did you bring the penis back that time?" he asked Pari. Pari said no, he hadn't, it had been too small.

"Oh, and down in Viseu they sell tapir penis for thirty cruzeiros. For old men," he explained to me, "old men who can't do it."

"Mundik's longing for one," said Chico maliciously.

Pari told me two odd myths about tapir. One was about the time when tapir were twice as large as they are today: a man crept up behind such a tapir and put his hand up its anus. The tapir, startled, bounded away and broke the man's arm off. Maír at once made the tapir both smaller and wilder, so that this should not happen again.

Many Indian myths look quite senseless at first sight, but a little comparative mythology is often a help in finding out their meaning. There is another version of this story current among the Indians of Guiana, which gives the reason for the man putting his hand up the tapir's anus; he thought he could get hold of the liver that way, without having to kill the tapir. For the tapir, in Guiana, is closely linked with the miraculous tree in the beginning of the world, which bore every kind of cultivated fruit, tuber and root. Only the tapir knew of it, and it would go and eat the fruit that fell from its branches. When other people discovered the tree they felled it and planted the manioc, the sweet potatoes, and all the other plants, but chased away the tapir because it had been selfish. Tapir's liver is a great delicacy, and the man who thought he could get it without killing the tapir was like the tapir itself, who was content to eat the fruit from the miraculous tree without felling it and cultivating the plants that it bore.

Pari's story was thus a very moral one: it implied that human beings, when they are selfish, are no better than animals, and that to become truly men they must cut themselves off from the state of nature the tapir represents. His second story showed this even better: it told how the first akang-apar turtle was made. (*Akang*, a head; *apar*, to bend. The turtle (*Platenys spyx*) is so called because when frightened it bends its neck and puts its head to one side of the shell, instead of retracting it inside.)

Akang-apar was originally an Indian who had some odd habits. One day when it was raining hard he went hunting and came back after a long time with a couple of tapir ribs. Only they weren't tapir ribs, they were his own. The next day it was raining again, he went hunting and came back with a tapir's tail—but it wasn't a tapir's tail, it was his own. His brother wanted to see how he managed to kill tapir when everyone else had failed, and he said, "Tomorrow, let's go hunting tapir!"

"No," said Akang-apar, "it's a long way off to where the tapir are."

But one day soon after he went, and his brother followed him secretly: the man went a long way off till he came to a small stream. There he cut some assaí palm-leaves, spread them on the ground and lay down on top of them. The brother crept up and watched from behind a tree. Then the ang, the soul of Akang-apar, appeared, took a knife and cut Akang-apar's arms off at the shoulder, first the right arm and then the left—aie! shrieked Akang-apar, aie, aie! But his ang took a tawari cigar* and blew

* These cigars are made of native-grown tobacco rolled up in a thin sheet of bark from the tawari tree. For their use in shamanism, see Chapter 15.

smoke on the wounds, and two new arms sprouted. Akang-apar's brother went home quickly and told what he'd seen. When Akang-apar returned with a pera on his back containing two tapir legs he called to his wife—"Wife, come and eat tapir!" But the wife said, "That's not tapir, that's your own flesh, no one wants to eat your flesh here! That's not tapir!"

"Yes, it's my own flesh," he said. And he went away, with his wife, till he came to a pool, and he jumped in and became an akang-apar turtle; his wife became a saracaca. And his two tapir legs, which he'd left lying on the ground they were eaten by other turtles, by akang-pinim, mashiré and pia-pen.

Selfish or self-sufficient, there is not much difference. The tapir is a self-sufficient animal in an obvious manner, for it is said that the female does not chase her son away from her when it's grown up, but commits incest with it; if it's a daughter, the two of them stay together and mate with the same male. So the tapir is a symbol of that self-perpetuating, self-devouring state of things, self-sufficient not in spite of Nature, like Maír their culture-hero, but because it is Nature herself. Akang-apar, who might have been a man, turned into a turtle because he was caught trying to be self-sufficient in the wrong way. The turtle is the lowest form of self-sufficient life, since it combines male and female symbols in one: the shell is female, the long neck and legs inside it, male.

But there are no taboos on the eating of tapir, except the general ones which apply during pregnancy and child-birth, or when Indians have to go into ritual seclusion. A pregnant woman mustn't eat tapir, for fear that her child will get an enormous head that will stop it being born; besides, it will make the man who shot it *panem*, unlucky. If she or her husband eat tapir before the child has been named, the child is likely to die and the parents to fall ill. But these taboos apply to all game.

We ate and ate the night Pari killed the tapir. It was as fat as a tapir should be, and therefore extremely appetising. Nothing tastes as boring as a lean tapir: after a few heavy mouthfuls you long for some juicy fat. Boiled tapir is dull, too, but boiling is the quickest way of cooking it. Tapir meat tastes best when it is cooked slowly on a moquem, a platform of cut saplings supported off the ground by forked posts. Pari prepared his moquem the next morning, and laid the meat on top without taking the skin off. He kept a low but steady fire going all day, so that by evening the meat was well cooked, faintly smoked and still juicy: the skin kept it from drying out. When it was cooked, he sent bits to Tahi and Taperé. Toí had also made a moquem for his share of the meat. His hut was too small to contain a moquem, so he built it outside and covered it with a small shelter to keep off the rain. The thatch of naja palm fronds dried out by the evening, when a spark set it on fire; there was a great rush of flames, and agonised cries from Paijé, who splashed water over it. The fire went out without spreading to Toí's hut, but Paijé had broken Terepik's best gourd bowl in trying to put it out, and Terepik was very angry.

All large game tastes better when it has been smoked, especially deer, which boiling is apt to make tough. There are special taboos surrounding deer meat, for deer is not sô'ô-té, real game like tapir and pig, but sô'ô-ran, false or pretence game. All through Brazil deer are thought of as having something to do with the ghosts of the dead, for they come and eat the growing manioc leaves in the roças and spend much of their time in the deserted capoeiras, where the dead lie buried. They have, too, the eerie habit of making a sudden noise when they're startled, crashing away in the undergrowth; three seconds later there's not a sound to be heard, they've gone.

The man who kills a deer may not bring it into the village himself. He lays down the pera containing the meat at the edge of the clearing, and sends his wife to get it or, if he has no wife, another woman, or even a man who has not been hunting that day. Certain pieces are kept aside to be boiled, while the rest is laid on a moquem and smoked. Deer meat must never be grilled over a naked fire, or the cook will get a terrible fever and get *kaú*, crazy, as would the man who brought his own game into the village. And when the meat is cooking you should not make love, put on your feather ornaments or paint yourself, or even sing—an Indian ticked me off properly one day when I was humming over a plate of deer meat, and said I'd get bald if I continued.

All these things are forbidden apparently in order that the deer should not remember that he too is a human being, for though the Indians were once cannibals they could only eat human flesh with special precautions, during particular ceremonies. The taboo on grilling deer meat is more puzzling, but is paralleled by similar taboos on the flesh of macaws, trumpeter birds and nhambu, or forest hens. Macaws, in addition, may not be cooked or eaten during the night, or trumpeter birds during the day, lest everyone in the village fall supernaturally ill. For the macaw is an emblem of the sun at noon, and the trumpeter bird of the sun at night, harmless as long as they are not insulted by being consumed with their own attributes. That would be equivalent to saying the fire over which they were grilled was really the sun, and that the man who was doing the grilling was the sun's owner, Maír. Such mythological identifications are possible only in rites; the man who makes them in daily life is apt to get burnt up in the power he has unleashed.

The nhambu is a special case since it, like the deer, is considered to be a spirit. But all birds have spiritual powers which make them dangerous at certain times, since they are Maír's emissaries. No child may eat bird flesh till he's about ten. And nor may dogs. Bitches, when they have puppies, are made to keep the same taboos as women in confinement; they are not taken hunting, they are kept on a diet of manioc flour and tortoise meat, for if they eat game not only is it thought their puppies will die, but the hunter who killed the game will become panem. Even when their 'confinement' comes to an end they may not eat birds, since a dog is a kind of child who never grows up. It can never become a man, but it is still human, since it eats human food.

Meat, whatever it is, has to be cooked thoroughly, or the Indians won't eat it—the slightest sign of redness inside, and back it goes to the fire. They are also suspicious of meat from strange animals like cows or domesticated pigs, for the spiritual qualities of such animals aren't known and might prove extremely harmful to those who have no proper relationship with them. They felt much the same way over the few tins of meat I'd brought with me as a change from the usual diet. Chico once gave a piece of spam to an Indian, who ate it and at once begged for another bit to give to his wife. He gave it her and urged her to eat it—he was going to die, he said, and he wanted to be kept company. She divided the morsel in two, and gave half to her young son who thrust a bit into his sister's mouth saying, "Now we'll all die together." We comforted them, saying it was only pig, but they were sure it was some horrible animal, and that anyhow it had been dead in its tin for months. It was good spam, however, and after waiting five minutes they asked for some more.

The dish they particularly liked was spaghetti, and I'd have to put a huge pot on the fire so that everyone could have some. I quickly ran out of it for I couldn't refuse without being thought extremely mean, food being something which has to be shared. Whatever game a man kills, he should only keep the head and the spine for himself, and give away the rest. The best pieces he gives to his nearest relations, his brothers, his parents-in-law, especially to his brothers-in-law, and if there is anything left over the others in the village get it. Usually there is enough for everyone, since the villages nowadays are small; and I certainly always did well out of the division, whoever the hunter was.

One doesn't divide birds before they're cooked, because there's so little of them. Nor are tortoises customarily divided up. The tortoise is a kind of tinned meat, since you can hang it up alive from the rafters for a week or so without it dying, though it's better if you build a small cage on the ground under a cashew tree perhaps, where it can feed on the fallen fruit. The tortoise represents one of the few forms of surplus property the Indians can accumulate. Pari would get a lot together from time to time, and carry them to a Brazilian township in order to exchange them for cloth or knives, though he was once rudely disappointed after carrying a dozen tortoises on his back to Chega Tudo, several days' walk away. He found a glut of tortoise there, so that no one wanted his, and he had to carry them all the way back again. He still didn't quite understand about money, for he told me that he'd managed to get some clothes there all the same, and when I asked what he gave in return he said 'nothing, only money'. (He had got it at Canindé.)

I never like seeing a tortoise being prepared for a meal. One can't kill a tortoise properly, and even if one could the Indians wouldn't bother. What you do is lay the tortoise down on its back and then start hacking away with a machete till you've cut the plastron or undershell free from the upper shell. The plastron is cut free from the flesh with a knife down the neck, and then torn away, revealing the tortoise pitiably still trying to hide its head and legs. The guts are ripped out, head and legs cut off, and the two empty shells placed bloody-side down on the embers to cook, for there is still a lot of fat meat on them.

The head and limbs, after being scraped clean of scales and dirt, go in the pot to boil. They are very tough, and it often takes four or five hours before they're ready to be eaten. But tortoise meat is good, slightly gelatinous; the only drawback is that if you eat tortoise too long without a change, you're liable to come out in spots.

The best parts of the tortoise are its liver and its eggs. The liver is often boiled with the eggs, but it is best grilled, oozing drops of fat into the fire, and then eaten with a lot of manioc flour and some eggs. It has a sweet delicate taste, and a silky texture, quite unlike the rather coarse and vulgar taste of cow's or sheep's liver.

There are other ways of cooking tortoise. I saw Mbeiju once with a small tortoise hardly four inches long, which she laid back-down on the ground. She picked up a thick stick and with this punched a hole in the plastron; she dragged the guts out of the hole, filled it up with manioc flour, and put the tortoise into the middle of the fire, still alive. Tero did the same with a small terrapin once, without even bothering to clean it—he just threw it into the flames, poked it back every time it tried to come out, and laughed.

CHAPTER EIGHT

The Hard and the Soft

IN the days when I was preparing to visit the Indians for the first time, I thought I might find it very hard to get on with them. How did one behave in an Indian village, what did one talk about? I asked myself. This fear, which was strengthened by hearing horrid tales of their savagery, was of course quite absurd: after being in their village for a week or so, the Indians lost their shyness and suspicion, I my natural stand-offishness, with the result that I felt as much at ease as if I were staying in some small English village—more so, perhaps, since I did not have to be respectable. There is no privacy among Indians (except in matters of ritual and magic) and they are, besides, the most tireless of gossips: start them off on any one of the three universal subjects of conversation, war, sex or religion, and they will at once show just how human they can be.

From time to time, however, the contrast between this undemanding domestic life and the hard, brutal conduct they assume in war struck me as odd. Their idea of what a chief should be seemed quite at variance with what he actually was in the daily life of the village; and the set idealism implied in their ne'eng hantan or hard speech, in their ceremonial feather ornaments and the way in which they painted themselves, hardly fitted in with the kindness and tolerance they extended to strangers such as myself. The difference is perfectly summed up by the Indians in two words, hantan and membek. To be hantan is to be hard and strong, to be a real man; to be membek is to be weak and soft like a woman. A man needs only a little opposition to change from an accommodating state of mind into a violently aggressive one: a change as startling as it can be unpleasant, even though a lot of his aggressiveness is a bluff which can often be called.

The jungle is a place which encourages this contrast. I had my first taste of its effect when walking through it with Ribeiro the year before. We had been met by João Carvalho, the officer in charge of the post at Canindé, who had brought along a number of Indians to take us through the jungle and to carry our baggage. I entered the jungle confidently enough, as though I were Bates, Wallace and Speke all rolled into one; how pleasant, I thought, to be an explorer! I changed my mind as I began to trip over roots and low-hanging lianas, though the feeling came back when we camped that night by a small stream, slung our hammocks between trees and roasted some peccary meat over a fire: it was only later, however, that the jungle began to affect me as I had once imagined it would, with a feeling of things unknown and dangerous surrounding me. This was when we came across the tracks of the nomad Indians called Guajaja, whom the Urubus hate and fear. The Guajajas have no machetes, so that to mark their trail through the jungle they would catch hold of saplings every ten or twenty yards, give a sharp twist and break them in two, instead of slashing them as Urubus and Brazilians do. Whenever we came upon such signs, often freshly made, our Indians would call a halt, slip off their loads, tighten the strings on their bows and go silently to one side of the trail hoping either to kill a Guajaja man or capture one of their women. After that, the jungle did not seem as empty and inconsiderable as it had before, nor did the idea of receiving an arrow in the back appear far-fetched.

No Indian can be free from this feeling of suspicion when he goes hunting, even if he stays near his village; and the feeling becomes much greater, of course, if he decides to go on a long journey to hunt some rare bird for its feathers, or just for the love of adventure. The jungle isolates, cuts off, closes in; to an outsider like myself it induces a kind of wary short-sightedness and a longing for large views and open spaces. To the Indian whose world it is, the jungle can hardly evoke the same feeling, for he can read its signs and understand its noises; yet to him also it has a definite presence, and he peoples it with ghosts and spirits that embody his apprehensions.

Ghosts, of course, cannot be killed, which is why they are so troublesome. A ghost is, as it were, a subject who has no ego: it can act only when it possesses a man's attention. There is a story of an Indian who did try to kill a ghost (it is told in Chapter 14), and the result was obvious: the man died, for he was being violent only to himself. Luckily for the Indians, however, the jungle contains other hostile inhabitants which can be shot at without supernatural forces coming into play. These are other Indians, belonging to enemy tribes. Listening to the Urubus talk about the past, it seemed to me that they were on the worst possible terms with every other tribe in the district; they were suspicious of all who were not one of themselves, and shot strangers on sight. They fought the Guajajas, because they were nomads and therefore like animals; they fought with the Tembés and the Timbiras because they were river Indians, and therefore fair game; and they fought with numbers of other tribes who lived in the jungle and who had the same ideas about strangers as the Urubus themselves. There were, for instance, the Terembé-hu or Big Lips, the Wira-hu-man or Ancient Hawks, and the Juru-pihun who got their names from their habit of painting a great black stripe over their mouths—they were remarkable too for being quite bald and for staining their scalps with red stripes, like a beach-ball. There were the Merikerns, who blew a fatal, magic dart from their mouths—presumably a description of some Gê tribe to the south, who have blow-pipes—and the Timiruku, living near the headwaters of the Gurupí: they are supposed to be pygmies, standing but four feet high, who walk about with great lances (muruku, hence their name) and a couple of clubs under each arm.

Perhaps the two most interesting tribes of which the Urubus speak are the Kuashimbir and the Andirambir. Kuashi is the name for the coati-mundi, a small raccoon-like animal; andira means bat; and -mbir is a suffix used whenever talking about someone who is dead. The Coati and the Bat people, therefore, are tribes of Indians who are quite conceivably ghosts as well. They are supposed to wear their hair in a top-knot, to live in round huts shaped like bee-hives and to sling their hammocks right up under the thatch, where they sleep during the day. They wake up at night, like bats, to work in their plantations or to go hunting. The Urubus, who fear these two tribes, presumably give them such demonic attributes because they know so little about them: a good example of what the jungle can do to a fearful imagination.*

The jungle thus forms the necessary contrast to the safe and friendly world of every village, and so makes sense out of that pair of opposed moral qualities, hardness and softness. Within the village, the Indian can afford to relax, but when he passes beyond the edge of the clearing, he steps out of the world in which human beings are protected by the life they have together, and enters the

* Both these tribes do in fact exist. The Kuashimbir are said to have enormous bows designed to be drawn when lying on the ground; the bow-stave rests on the sole of the foot, the bow string is grasped by both hands and the bow is opened by straightening out the leg and flexing the arms. I have seen what must have been such a bow in Rio de Janeiro, for it was quite four inches wide; unfortunately I have forgotten where it came from. The Andirambir, curiously enough, seem to be the Urubus themselves, or at least their *tushau* who, after killing a man ritually, goes into seclusion for five or six months, resting all day in his hammock which is slung high up under the thatch of his hut.

jungle world that can be dealt with only on its own terms. It is an unmoral world, in which the same thicket can be used either as a hiding place or an ambush: the only positive virtue it requires from a man is that he be on the alert. But, just because moral values have no place in it, the Indian cannot think of the jungle as being a neutral world: it does not relieve him from moral responsibility, but threatens his existence. The jungle spirits which the Indians speak of all have the same general characteristics: they induce panic and try to bring a man down to chaos. The simplest way they have of doing this is to make him lose his way and to wander about aimlessly looking for some familiar landmark. This experience, unsettling enough, can easily turn into something much more serious, for the Indians sometimes lose their way metaphorically as well as literally. The story of the man who got mixed up with a lot of Anyang ghosts, and eventually turned into one of them himself (Chapter 14) is an awful example of what may happen.

Alertness, then, is the quality most needed if a man is not to lose his wits or his way in the jungle. 'I am hard, I don't need sleep' is a common boast among them, just as it was among the old Tupinamba, who called a warrior a 'kerimbae', one who does not sleep. To sleep is to die, both in a literal and a moral sense, as the Urubu myth of the first created man tells. There is a tree in the jungle called wira-pirok, the peeling tree, which is thought to be as immortal as Maír, since it continually sheds its old skin like a snake and becomes new again. This tree, the myth says, once started moaning loudly in the middle of the night—'wooooooooo!' it moaned, its voice rising and falling; and every creature that heard it immediately went towards it, and became immortal. These creatures included the spider, the cicada, the snake, the scorpion: all animals that shed their skins and can therefore rejuvenate themselves. But the first man was lazy and took no notice of the moaning tree; he went on sleeping, and so brought death into the world for human beings.

The Indians' continual talk about hardness in the end sounds a bit forced, for they insist on it too much. Presumably this is because they not only fear being *membek* or soft, but in a way really are soft. This is not to say that they are weak or cowardly, though a coward would with justice be called a soft man; the true meaning of the word seems to point to a kind of moral laziness. To be soft means to be irresponsible, shiftless, to have no large purpose in life; and in these terms all Indians are indeed soft for a great part of their lives. They will moon about the village for days if they feel like it, sitting in each others' hammocks and idling away the time with stale gossip, and they will do this even if there is no food in the village but manioc flour—which, though excellent when taken as a shibé to fill in the gaps between meals, is not satisfying as a main course. Then, perhaps shamed into it by their wives, they will reach for their bows and arrows and go hunting; and when they return in the afternoon with a fat deer slung over their shoulders, or a leaf knapsack full of tapir meat, the village once more will become active, and the men will think of building themselves a new hut or making a new bow-stave. Activity and laziness are both catching, but it is hard to be active when nothing makes a definite call on your energies, when nothing new happens for weeks on end. The amount of spare time an Indian has on his hands is enormous but, as one might expect, he makes little use of it.

The great importance the Indians put upon hardness as an ideal state is partly due to this intermittent debauch of laziness in which they indulge. To some extent, hardness represents the other extreme of impulsive action, but these impulses are only effective if they are disciplined. Hardness is a virtue because it implies responsibility and a purpose, which is why the chief always used to wear his feather ornaments in the village, since he is a minor Maír; he also had to remind himself of his responsibilities, as it were, by sleeping with his wife only once a month, instead of as often as he pleased. Hardness also has a sexual meaning. But there is one act in which self-interest and self-discipline are identical, and that is the act of revenge; for there inflamed impulses are made to work in the service of the ideal.

The curious mixture of licence and principle which one can see in Indians is often difficult to understand, and it is no wonder that they are fearfully known as savages: once they are in action, their motives are invisible. The Indians, of course, find civilised motives just as incomprehensible. Seeing the complicated world around them, their own is made to look small and petty: inevitably, they try to imitate the ways of the Brazilians, without understanding the values lying behind them, at the same time treating their own with a certain scorn. Nearly all Indian tribes, once pacified, succumb to this temptation; falling between the two stools of their own morality, which they have discarded, and the morality of the Brazilians, which they do not understand, they end up in a kind of limbo and after a few generations die from the despair of not being anything in particular. The various Timbira tribes along the river Gurupí have all vanished as tribes, for the handful of Timbiras who still exist have managed to do so by living the life of a Brazilian peasant, and discarding their Indian traditions. The few Tembé Indian villages that continue to thrive do so because they go on practising their traditional shamanism, that intercourse with the spirits that lies at the heart of their morality; but even they are finding life difficult, and are retreating to the headwaters of the Gurupí where civilisation cannot reach them. One cannot help feeling that the Urubus, whose spiritual life is not fostered by a native shamanism like the Tembés, are doomed, and that in a few generations the tribe will have disintegrated, and its surviving members have turned into rather sorry peasants.

The same fate overtook the Tupinamba, when European settlers became too numerous on the Brazilian coast. They did not, however, succumb without protest: their last years were marked by great messianic movements, in which thousands of Indians followed shamans in an attempt to reach the Land without Evil, where the golden age still lasted. (Similar movements, inspired by despair, occurred among the Plains Indians in North America: revolting from the world of white civilisation, they went on a crusade inspired by their old mythology.)

Reading the accounts of the old travellers, one can even see the point where the tribal life of the Tupinamba began to disintegrate; it was when the Portuguese stopped them from making war on their neighbours. This outraged all the young warriors, who were not thought men, and could not marry, till they had killed an enemy; but they got round the prohibition at first by raiding the cemeteries of their enemies, where they dug up the skulls of the newly dead and broke them to pieces with their clubs. This served in lieu of a ritual execution, where a living prisoner is clubbed to death, but the substitution was not really effective. The ritual allowed a youth to enter into manhood, but he still had to prove his quality, which he could only do in war, where Indian morality expressed itself so fully. Without war, the Tupinamba had no way of justifying themselves through their traditions.

The same facts unhappily apply to many Indian tribes: war seems to be a condition of their existence. The Urubus no longer ritually execute prisoners, and the young men, now that raiding has very nearly ceased, are no longer compelled to exercise their full alertness; so they treat the ideal of a warrior with a certain carelessness, or else, having no opportunities to be warriors, they invent adventures in order to get prestige for themselves.

Such an attempt to get prestige happened during my first visit among the Urubu. A young man called Yapo alarmed his village one evening by seeing what he took for a Guajaja Indian lurking behind a banana palm near one of the huts; he immediately shot off two iron-headed war-arrows, or takwaras, both of which hit the banana palm plumb in the middle. This was not good shooting, unless the palm had been his target. But then a woman thought she saw a Guajaja on the other side of the village, among the bushes, and this so alarmed everyone that they gathered together all the resin they could find, made torches with it and set off for Canindé, some forty minutes' march away, shouting as they went to frighten off the Guajajas. Yapo went back the next day, searched the neighbouring jungle, walked along a path and was jumped on (so he said) by a Guajaja in ambush, whose arrow missed him at the range of four yards. Yapo fled and, while turning to get in a Parthian shot, cut his bow string on the sharp edge of his arrow. I thought it an unlikely story.

The news was brought to the village where I was staying by a succession of visitors. With each telling of the story, the atmosphere in the village became tenser and more excitable, and the story itself became grossly exaggerated: the number of attacking Guajajas grew till there was in the end a force of some twenty, who were supposed to have beleaguered Yapo's village; and these men, originally described as being of normal size, though large and strong, in the end seemed real ogres, seven feet high, of incredible strength and agility, accompanied by fierce black dogs, and armed with vast bows and *taboca* arrows whose leaf-shaped bamboo heads were a yard long. One evening, when the last instalment of this tale was being told by two visitors, a whistle was heard on the path leading to our village, immediately interpreted as coming from a war-party of Guajajas: the men flew to their bows and arrows and went out along the paths to see what was up, the women started shouting to their young sons to go too, and all the dogs in the village started barking and howling: the non-combatants assembled near my hut, and urged me to take down my ·22 rifle. It wasn't long before the men came back, slightly shame-faced: there were no Guajajas anywhere, and the whistle must have come from an owl. Indeed, I heard an owl soon after, whistling derisively.

When I decided to go to Canindé to see what had really happened, I found myself in the middle of a kind of war-party, all anxious to go and kill Guajajas. We met Anakãpuku on the path, who had had the same idea. He had a great bundle of arrows, and was painted with black genipapo in stripes and circles over his face and body, and he carried a miniature jamashi or back-basket made of woven split lianas, in which was his hammock and some food. When we approached Canindé, Anakãpuku suddenly warned me not to be frightened: they were all going to yell. "It's awful," he said, "you'll be afraid!" Yell they did, a moment later: a high, long-drawn rusty whoop, rising and falling, a kind of moo in the top tenor reaches. They did this, they explained to me smiling, to warn all their friends that they were on the war-path. It was a horribly disturbing noise.

At Canindé the regional inspector of the Indian Service happened to be paying a visit, and he tried to calm the party down: the Indians that Yapo had seen were not Guajajas, he said, they were friendly Indians from the Capím river. Anakãpuku at once went into savage tantrums—"No!" he bellowed, with terrific insistence, "they're Maku!"—another tribe, reputed to be even more ferocious than the Guajajas—and he dropped his bow and arrows with a clatter, seized his left wrist with his right hand and fiercely, ravenously bit it: bit at it, rather, because all he wanted was to show that the Maku were cannibals who ate people alive. I could almost believe it, seeing his wet, yellow teeth and his eyes screwed up with rage.

But even his vindictive determination to hunt down these Maku—or Guajaja—in the end came to nothing, for the jungle was quite empty of them: he and the other Indians spent two days searching there, and did not find as much as a footprint.

The whole incident made me curious, and I began to wonder if Yapo could have had any special motive for inventing the story of his private and inconclusive battle. There were, indeed, some interesting facts which I later found out. The chief of the village was a blood-thirsty man called Ingarassu, who had a record of making trouble at Canindé, and who had delighted in various savageries in the days before the pacification. He, at any rate, still believed in being hantan, and bore himself with a defiant air of bravado. His boasts and the stories of his exploits must have galled Yapo exceedingly, and provoked him into bolstering up his self-respect in some way: which he did by shooting at his imaginary Guajajas. But Yapo had a further reason for his play-acting. In the old days, when an Indian died, the survivors would get so angry that they went to take it out on the Guajajas, thus pacifying the ghost of the dead and assuaging the grief of the living. Yapo's mother had died a few months before and, though her death was from natural causes, it may have lain heavily on Yapo's conscience: so that his play-acting was not only meant to get himself prestige, but to honour her ghost. In the end, however, it did him little good, for most of the other Indians concluded, as I had done, that he had been lying.

There are, however, a number of enemy Indians who still live in the jungle nearby, and should an Urubu meet one of them arrows are certain to fly. A classic story of such an encounter was that concerning Anakãpuku, who had been wounded in the shoulder by a Guajaja and in revenge wiped out an entire Guajaja village. This was one of the first stories that Antonio-hu told me, and I enjoyed it thoroughly. Antonio-hu was a barrel-chested man, full of a great and boyish strength; he was getting on for his forties, and was already a bit deaf, so that he would shout sometimes without meaning to. These shouts were always full of good humour, because Antonio-hu enjoyed everything he did, even when he was sad: and he could be sad about the smallest thing in such a way that one wanted to laugh with pleasure. His good spirits were completed by a charming giggle, somehow not incongruous with his size.

Antonio-hu told me the story to show me how tough a man can be. In spite of our friendship, he liked to sneer at my courage from time to time, though in a brotherly way, and he said that I would have been scared stiff even at the sight of one of the old warriors, of a tushau in his maracaja-skin cap, painted for war. I was quite glad not to be put to the test, and as his jeers were only a harmless way of getting his own back at me for being a white man, I didn't mind.

Anakãpuku, he said, once went on a trip to the Pindaré river to shoot cigana birds (*Opisthocomus* sp.) for their feathers. He had built himself a small shelter there, and one day when he was out hunting some Guajajas came and took everything he had left in it—his hammock, his spare machete, his clothes, his manioc flour. When he returned and discovered the theft he became angry, very angry indeed, and next morning he set out after the Guajajas. He was all puffed up with rage, said Antonio-hu, he shuffled along with both arms flexed tight against his chest, his mouth an angry shape with the lips protruding, while his breath came heavily, spaced out with furious grunts. The Guajajas had left tracks everywhere, and Anakãpuku followed them easily enough; but suddenly a single Guajaja stepped out from behind a tree and shot a taboca at him, a bamboo-headed arrow. Anakãpuku saw it just in time: he ducked, and the taboca got him, not in the chest where the Guajaja had aimed it, but in the upper arm. It came out behind this shoulder.

"Ah!" cried the Guajaja, "now I've killed you!" But he hadn't: Anakãpuku was in a fury, and though the taboca stuck out of his shoulder he made after the Guajaja. Antonio-hu demonstrated the frightful tottering shamble of Anakãpuku trying to run after the Guajaja, who fled before him. "The blood!" said Antonio, "the blood that came out of him! like when you break open a tortoise near a stream, and the blood flows into the water."

But Anakãpuku got back home all right, swearing he would be revenged. Next year, when his wound had healed, he took a large party of men with him from various villages and led them to the place where he'd been shot. There they saw paths with many footprints, footprints of men, and women, and many children. "I'll go along this path!" said Anakãpuku, and he took four men with him, and the other men went on other paths; on each path were five men, all hunched up with rage and grunting to themselves, till they came to the village. There was an old man, an old Guajaja with a beard—but a beard, said Antonio-hu, it came right out all round his face!—and he was dandling a baby, throwing it up and catching it, throwing it up and catching it, while his wife looked on and smiled. Then she happened to turn round and she caught sight of Anakãpuku. She thought he was just another Guajaja come to visit so she laughed at him, ha ha hee! But Anakãpuku was in a rage, he'd spent a month making nothing but tabocas, sharpening them, making tabocas and sharpening them. "Don't let the men get to their arrows!" he'd told everyone, "or we won't be able to kill them all." There was the woman laughing at him, but he opened his bow and shot . . . wahhh, the old Guajaja was dead, an arrow in his chest; the woman tumbled out of her hammock, and there was an arrow up her arse; and the other Guajajas who went for their arrows beside their hammocks (for they always keep their arrows there) were all shot. Anakãpuku shot seven, men and women; everyone in the village was killed, including the children. Anakãpuku was angry, because his wound was still hurting.

It did not strike me till afterwards, so carried away was I by Antonio-hu's pantomime—for whenever he described a battle he would leap out of his hammock and charge about the hut loosing imaginary arrows—what an improbable story he had made of it. The villains of the piece can hardly have been Guajajas, since they have no settled villages, and since an Urubu cannot understand their language. Yet Antonio-hu at once followed up the story with a little vignette of a Guajaja woman in the village calling out to Anakãpuku and begging him not to shoot her, but to take her away with him instead. "Don't shoot," she was supposed to have said, "cararan putar! I want your penis!" Cararan is a good Urubu word (as is putar, to want), analogous with the name for the vagina, carapua: cara being the name of the red sweet potato, cararan being that which looks like a potato, and cara-pua the potato with the round shape. Presumably Antonio-hu was embroidering a plain tale with fancy trimmings, as his custom was; an embroidery which showed, however, how he thought Indians should behave.

So what he made the Guajaja woman say was the right thing, indeed the only thing that might have saved her life: for if some Indians go on raids inflamed with bloodlust, like Anakãpuku, others go in the hope of getting a woman for themselves. It is not just because there are not enough women in the tribe—for a man is really only contented when he has two wives—but that women from other tribes, especially Guajaja women, are thought to be especially pretty. Their skin is whiter, their hair finer, their noses straighter—or so the Urubus say—and they wear necklaces of chopped up piripiriwa root, which gives them a delicious smell. Antonio-hu was always going on about Guajaja women, like a Londoner describing the joys to be had in Boulogne, and he hoped that one day he might come across one in the jungle and catch her.

"But Antonio-hu," I asked him once, when he'd been maundering on in this vein for a quarter of an hour, "doesn't a Guajaja woman run away when you catch her?"

"*No!*" he said firmly, "they only want your rankuaĩ"—leaning over and putting a hand between my legs—"that's what they want, to feed their carapuas! You make a bed of palm leaves in the jungle, the woman lies down on it, and there you are!" He sat back in his hammock and viewed me complacently. Dear Antonio-hu, I thought, always making life appear so simple.

In fact, the stealing of a woman is not an easy matter, and a man needs all his hardness to catch her, bring her home and tame her. I was told of a raid that happened some years ago, when two or three Urubus out hunting came across a Guajaja encampment. They hid in the bushes to start with, watching the Guajajas as they went about their business—they were especially fascinated at the sight of a Guajaja man having a piss, because he had no string tied round his penis, as the Urubus have, but a piece of bees-wax which he stuck on over the end. When, at last, the Guajaja men left the camp to go hunting, the Urubus jumped up and ran in: the Guajaja women shrieked, slipped off their skirts (which, from all accounts, are woven of tucum palm fibre and are worn like a sari) and ran naked into the jungle. The Urubus followed them and managed to catch a lusty young woman who shouted in fury and bit them as they took hold of her, but they just laughed and said, "No, no, we're good! Don't bite, we're taking you home with us, So-and-so's going to marry you!" And So-and-so, who was there, then became hantan in another fashion, stroking her and feeling between her legs, he felt so passionate. They dragged her home, gave her a hammock to sleep in, and sat round all night watching her. "Don't try to escape," they said, "you can't escape, we're going to keep awake all night, looking at you. We don't need sleep, we're hard!" Then, after a few angry and bewildered nights, the woman agreed to stay: she got a liking for manioc flour and for cauin or beer, neither of which the Guajajas know about, she was given some beads and a cotton skirt, and her ears were pierced so that she could wear ear-pendants.

These raids to steal women are more important to the Indians than at first sight appears. Indeed, the Urubu tribe owes its very existence to a couple of Guajaja women stolen some five generations back, which may be one reason why such women are so highly thought of. In those days, the Brazilian settlers were beginning to push their way into the jungle, with the unhappiest effects: for the Indians who lived close by were forced to move and to settle in the territory of other tribes, who naturally disputed their right to be there. The Urubus seem to have been caught in the middle of these movements, and lived in such fear of raids, both from other Indian tribes and from the advancing Brazilians, that they took to the woods like the Guajajas, and neither built permanent villages nor made plantations for manioc. They wandered about, living on game, honey and a flour made from arapari and anawira nuts, and looking for a part of the jungle where they could once more settle down in peace and live their old life. This, at any rate, is how I piece the story out, for the Urubus are ashamed of this part of their history, and do not like to talk about it: it is all right to steal a Guajaja woman and marry her, but to live like a Guajaja is to be on a level with animals and ghosts, and to lose all self-respect.

It must have been at this time that two brothers got

separated from the rest of their families, made their way south across the Gurupí and there started to build up a new life. They raided a Guajaja settlement and carried off three women; they built a small village for themselves, made a plantation, and began the long job of raising their families. In the beginning, they killed all the daughters their wives bore to them, because girls are only good for housework and to marry off to friends; they wanted sons, whom they could bring up to be warriors, to help fight against their enemies. They had five sons each, who when they were grown up, all stole Guajaja women and married them; and from this stock the present Urubu tribe is descended.

I have had to summarise the story for it is abominably long: Antonio-hu, when he told it me, took the best part of two afternoons to come to an end. But even this brief account of it is worth retelling, for it shows well how fear of the jungle breeds hardness, and hardness makes men. Fear of other Indians at first turned the Urubus into nomads without villages or manioc plantations, without even wives; they had become nonentities who were less than men. It was only by the most determined and long-drawn effort that they founded a village and raised a family, hardening their hearts even towards their own children in order to re-establish the old traditions which allow softness to exist in its proper place as the creator of domestic virtues, and not as their destroyer.

CHAPTER NINE

Quarrels

When the first Portuguese arrived in Brazil, they noticed that the letters F and L were missing from the Tupinamba language, and for some reason thought that the letter R was missing as well, though this was not so. From this interesting observation they concluded that the Indians were ignorant of the three institutions whose Portuguese names begin with those letters: the Tupinamba, it was declared, had 'nem fei, nem lei, nem rei'—neither faith, law nor king. The conclusion was that they were savages.

Such a point of view was perfectly justifiable. The Tupinamba were not troubled by faith, in the ordinary sense, since the supernatural world was for them a commonplace that could not be doubted; besides, their ceremonies were concerned not with the worship of a god, but with the celebration of heroic man. They had no king: for, though they had paramount chiefs, these were shamans whose power rested in intimidatory magic and not in an accepted idea of kingship: as for the village chiefs, they had position but no real power. There was no law: that is, no system of punishments for various offences.

These facts apply equally well to the Urubu. The ideas of 'king' and 'law' are of course connected: punishments are only possible when someone is in a position to see that they are carried out. But this does not mean that Indians are not moral animals: they have morality aplenty and if anything think too much of it, and justify their most unpleasant acts in its name. They consider a man is virtuous not when he is 'good', in our sense of the word, but when he possesses the power in virtue of which he is truly a man: when, in fact, he is hard. If, however, every man justifies himself by the tribal virtue of hardness, and there is no paramount chief to see that justice is done, what happens to law and order? This is what troubled the Portuguese, and made them say that the Tupinamba were only savages.

In such a society, the only way of dealing with a malefactor is to shame him into conformity, as a child is shamed when he misbehaves. There is really no problem of law and order within a village, since its members are nearly always related to each other: a man who takes to stealing, or who beats his wife, is subject to all the enormous disapproval which an outraged family can bring to bear on one of its members. This sense of family, indeed, stretches right through the tribe: a man will have a brother living in one village, a cousin in a second, an uncle in a third, his parents-in-law in a fourth, and so on, so that through them and the relations they make through marriage, he could probably find a relative in any village he might visit. Besides this, there is an obvious spiritual kinship between tribesmen that one can see working through their chiefs. Chiefs think of each other as brothers, even if they may be only cousins four times removed, or more; through this spiritual brotherhood their villagers are inevitably related.

This system of moral constraint on the whole works very well. It was true that Anakãpuku complained bitterly of people who broke into his hut and stole the machetes he'd acquired from Canindé, but that was because the situation was completely new: wealth in that form had never existed before, and traditional morality did not know how to cope with it. I myself found the Indians extremely honest. I never bothered to lock my suitcase, even though everyone knew it was full of beads and knives, for not only was I their guest and therefore under their protection, but I was not keeping my wealth to myself: I gave it away bit by bit, as morality required. The only thing I ever missed was a couple of pounds of sugar. This happened when the cupu trees were all bearing, and the Indians were making a kind of sour lemonade from the fruit; I would give them sugar to sweeten it. In the middle of the cupu season I went to another village for a couple of days, leaving all my things behind, which proved too much of a temptation for one Indian who had a taste not only for cupu but for sugar. I didn't mind very much when I discovered the theft, though the other Indians shook their heads gravely in disapproval; this made the thief so uneasy that he didn't come to talk to me for three or four days.

It was about that time that Toí discovered someone had stolen the beads I had given him. The day before, Antonio-hu had come over from his village to see me, dressed as befits a chief: he had on his best blue trousers, a somewhat dirty shirt, and a bright red cap crammed on his head. With him came most of the others who lived in his village, all expecting that I would open my suitcase and give them nice presents. I was not, however, feeling in a very good mood that day: I had a cold coming on, my legs were covered with sores where I'd scratched at tick-bites, the heat was somewhat oppressive, and I felt like having a short sleep. Antonio-hu, however, made this impossible, for he sat down beside me in my hammock, and began to talk. After a little he came to the point, though in the most courteous fashion imaginable. "My daughter," he said, pointing to her with a thrust of his lower lip, "my daughter wants some beads. She wants them badly; she hasn't got any at all." I laughed, which was as good a way as any of refusing, and said that I was going to visit him shortly, and that I'd bring them all presents then. Antonio-hu didn't insist, but I could see he was rather sad, and the women who had come with him were very disappointed. One of these was a tall, lanky creature called Curumiú, who looked a bit of a sloven, and the disappointment must have been too much for her. Beads were on her mind, and beads she determined to get. She went down to the stream, as if to have a bathe, waded along till she came to another path that led to Toí's hut, and followed it carefully till she reached the edge of the clearing. (This, at any rate, is how Toí reconstructed the scene: he went over the ground later and found her footprints.) Then she got down on all fours and scuttled along behind fallen tree-trunks till she'd reached Toí's hut and there found his store of beads, which he kept in a tin hanging from the rafters. She opened the tin, took out the beads, dropping some of them on the ground as she did so, and went back the way she'd come without anyone having remarked anything.

"What are you going to do about it?" I asked Toí. Toí looked mournful. "What can I do?" he asked. "The beads are gone." And when I pressed him all he said was, that if ever he had the chance of taking something of Curumiú's when she wasn't looking, he might; and again he might not.

Toí was unlucky, for he was to have another lot of beads stolen a month or so afterwards. The thief this time was his own sister, the mother of Tahi, or so it was suspected; one day, when everyone had left the village to go hunting, or to work in the manioc field, she went round each hut stealing all the beads she could find. Neither Toí nor anyone else did anything to get their beads back, but just went round complaining. As they talked they remembered other thefts that she had committed and saw in them the cause of her prolonged illness, her slight madness and her gonorrhoea. Her sins, they concluded, had found her out, and her latest piece of stealing would doubtless be followed by an even more severe act of supernatural retribution.

An affair like this leaves a bad taste in the mouth, for it breeds resentments that are hardly ever brought into the open. Theft is not considered to be a very grave offence, but there are other matters, such as personal slights, over which the Indians brood till a first-class hatred is hatched out, and whole families are ranged against each other. There are two sure ways of judging whether people are on good terms with one another. One is to see if they want to live in the same village; the other, if they have no fears of getting drunk together.

Cauin, or manioc beer, plays a curiously important rôle in Indian life. The process of making it is surrounded with various taboos, which by itself is enough to show that a magical transformation is taking place and that its outcome is in doubt. It is one of the earliest forms of alchemical working, where brute matter is made to turn into spirit: in Europe liqueurs represent the final stage of this transformation, being the quintessential distillation of herbs into an elixir—the pun on the word 'spirit' makes this clear. The Indians likewise believe that alcohol is a spirit, which possesses them when they are drunk. The possession makes them *kaú*,* crazy, a state in which it is recognised they are not responsible for their actions. It is only a step further to see cauin as a kind of spiritual cleanser, washing worldly preoccupations out of one's soul, and it is for this reason that huge pots of beer are always provided at ceremonies, the Indians getting drunk before they begin their rites. They start drinking as the sun sets, and go on all through the night, singing as they do in praise of the cauin: singing about manioc tubers pounded in the trough, manioc mash put through the sieve, manioc made into pancakes and baked on the oven, manioc pancakes fermenting in pots into cauin. At dawn

* Hence the name for beer, kaú-i, literally meaning crazy water. I use the form cauin by which it is best known in the anthropological literature.

they fall silent, and the women come round telling everyone that the sun is rising, and asking them if they got properly drunk the night before. Thus purified, they begin their ceremonies.

These feasts are the great occasions in an Indian's life, and visitors will come from miles around in order to attend one. Drink of course is not only a purifier but a begetter of conviviality: these two states are indeed hardly distinguishable from each other. A beer-feast is therefore the obvious way of bringing enemies together and making them settle their quarrels, and this seems to be how the Urubus made peace in the old days with other tribes. Tribal wars are always extremely difficult to stop, partly because there is no simple way of reaching agreement: there are no leaders who have power to treat with one another. Peace can only be reached through a personal agreement between every member of each tribe: they have to meet each other face to face and somehow turn their common hostility into friendship. They do this by getting drunk together, by dancing together and learning each other's songs, and often they end up by exchanging their bows and arrows or their feather headdresses. Peace demands as active a relationship between people as war.

But if quarrels can get settled during feasts, they can also get started there, because for a man to forget himself in drink often means that he loses his self-control. The only instance I ever heard of one Urubu killing another happened when a man got wildly drunk during a feast, picked up a machete and whirled it around his head in a sudden outburst of animal spirits; then, drunk as much with a feeling of power as from cauin, he rushed into a hut slashing at house-posts as he went. Unfortunately for him, the second house-post he struck at was really a man, whom he killed outright. The feast immediately broke up in disorder, the women running into the jungle with loud shrieks, only returning when the drunk had come to himself. No one did anything about the incident, for they recognised that a drunken man is not responsible for his actions; even if he had known what he was doing, I was told, revenge was out of the question, for how could an Indian kill his brother?

Because they know that, when drunk, they lose their self-control, Indians don't go to feasts where they could meet someone they have quarrelled with: not because they are afraid of fighting, but because the outcome might be tragic. Two such enemies were Anakãpuku and Caro, who took good care not to drink in each other's company. The matter was especially serious for them, since both were chiefs and therefore brothers, and for two chiefs to fight would have been a first-class scandal—their business is to put down brawls, not to start them. Besides, a chief's person is sacred, and a supernatural retribution swiftly comes on to the man who strikes one. But why they had quarrelled I could not discover.

Caro had also quarrelled with Toí, over Toí's sister, that same unprepossessing woman who stole and who was kaú in the sense of being loony and morose. She had been married to a man in Caro's village, and when he died decided to go and live in her brother's village, where Tahi, her son by a previous marriage, also lived. This made Caro angry, for he wanted her to marry his youngest son —there are never enough women in the world, and a young man will marry a woman old enough to be his mother if he can't get anything better. But Toí's sister didn't like the idea, and nor did Toí himself, and Caro was left to sulk by himself.

A few months later, just before I arrived at Pari's, Caro decided to make a beer out of cashew fruit. The wild cashew trees were all bearing, and his own little orchard of domesticated cashews, whose fruit are three or four times the size of the wild ones, were doing well also. He filled a couple of large pots with cashew-fruit juice and waited for it to ferment. The news got around, and when the beer was ready people from many villages had arrived at Caro's for the party, just to get drunk: no ceremony was going to be held. Toí felt slightly dubious about going, but Paijé, his mother-in-law, was determined not to be left out of things and, as usual, she had her own way: Toí had to go, taking her, Terepik his wife, and his two daughters. They had a lovely time. "I was kaú," Terepik told me happily. She was spinning some cotton when I asked her about the affair; she put it in her hammock and then dropped down on to her knees in the dust, where she rested her arms upon a tree-stump, and her head on her arms. There she stayed, her full breasts shaking loosely as she lurched from side to side. "That's how drunk I was!" she said, with a giggle.

Everyone got drunk, including Caro. Towards morning, when the others were happily singing in their hammocks, he started up suddenly, stamped the ground in a rage and shouted 'hu! hu! hu-hu!' There was a chief from another village present, who tried to calm him down —"Go back to your hammock, you're drunk; we're happy, we're singing. Go back to your hammock and sleep." But Caro took no notice, he just went on stamping and shouting, till he rushed over to where Toí was sitting, pulled the red cap he wore from his head, and struck at him with a machete. One blow got him behind the right ear, Toí told me gravely, touching the spot with a finger; one on the nape of the neck, one below the left ear, and another over the heart. Luckily for him, Caro only used the flat of the machete blade, so that he wasn't hurt much; luckily, too, all the other Indians immediately threw themselves on top of Caro and held him down, while the women and children ran off into the jungle. Half an hour later Caro was fairly calm again. Paijé bustled up to him, trying to put a good face on things—"Well, I'm staying to finish your cauin," she said. "When you make another

I'll come again." Caro wasn't having any of that. "No, no," he said in a grumbly voice, "I'm angry with you all, don't stay here, no." So Toí went, taking his disgruntled mother-in-law with him.

"So you're angry with Caro?" I asked Toí, when he'd finished his story.

"No," he said, "I'm not angry. I don't want to fight with him. He only fought me because he was drunk. But I won't go to his village again, I'll never go there to drink cauin any more. Instead we'll make a cauin here, from cashew fruit or bananas, we'll all get drunk here."

It was just as well Toí didn't feel angry. He was a small man, intelligent and a good hunter, but I suspect he would have been no match for Caro, a tall sinewy man who bore himself with a kind of gloomy splendour, and who continued angry for a long time. He passed near Pari's village shortly after I had arrived there, and was seen by Tero who was making flour in the manioc hut. "Where are you going?" Tero called out, politely enough. Caro said not a word but took a brand from the fire under the manioc oven, and went off down the path to spend the night in the woods. "He didn't even want to come and see you," Tero told me afterwards. "I said to him, 'Francis is here, he's brought lots of beads, come and see him!' but he just went away."

This was by no means the end of the affair. Two months later, Caro was in a fury again, first with João Carvalho, the young Brazilian in charge of the Indian post at Canindé, and then once more with Toí. It happened this way. Caro's elder son had been ill for many months, and had steadily got worse in spite of everything that had been done for him. In the end Caro decided the only thing to do was to get him to Canindé, where he could be dosed with pills and given injections, a medicine the Indians think extremely highly of. So Caro's son was brought there, carried all the way on someone's back, being too weak to stand up by himself; and, since João was in Belém at the time, they left the invalid in the hands of Chic' Ourive, whose reputation as a bush doctor had spread among the Indians. Chic' Ourive did what he could. He put the sick man in the driest corner of his hut—he hadn't looked after the roof for years, and it leaked abominably during the rains—fed him, and tried various herbal remedies, none of which were much good. On his return from Belém, João tried his hand and gave the patient, now extremely thin, some injections of streptomycin, but in vain; the man not only had some kind of bronchitis but malaria and, I think, T.B. as well, and he died some weeks later.

When Caro heard his son had died he was furious and, like a good father, determined that his death should be avenged. It was easy to discover who the guilty party was: Caro's son had died after receiving an injection from João, and it was therefore João who had killed him. So he set out from his village with four other men, bound for Canindé with murder in his heart.

His path took him through Pari's village. I had unfortunately left for Canindé just before, without knowing what was afoot, and the next bit of the story I had to get at second hand from Pari. Caro, it seems, came into Pari's village in the evening with black looks for everyone, his face set in a hard mask. "I'm going to make a peccary out of João!" he cried, "I'm going to fill him with arrows for having killed my son!" His sudden appearance filled everyone with terror, the women bolted incontinently into the jungle and the young men sidled off without calling attention to themselves. There were, however, two brave people in the village: Piahu, the youngest brother of Antonio-hu, who had settled down there for a few weeks; and Paijé, Toí's mother-in-law. Piahu went up to Caro as he paced round the village, fuming and trembling with anger—"You know," he said, "there are rifles at Canindé, and a rifle bullet goes a long way. When I was there, I saw João shoot his rifle from his house, and the bullet hit the rapids at the bend of the river. You look out! If you go there with an arrow, people will bring out their rifles, and perhaps it won't be João who gets killed."

Caro went on shouting, though a little less violently: his anger was a matter of honour to him, and not lightly to be forgotten. Piahu went on talking, using the hardest words he dared, for he understood that Caro wanted to be convinced, but had to be convinced without losing too much self-respect. Then Paijé added a few words, reminding him what a good friend João really was, and in the end he calmed down and stopped shouting, slung his hammock and sat down moodily. Piahu and Paijé went to their own huts, but kept themselves awake by smoking great tawari cigars, and by rocking themselves in their hammocks; everyone else spent the night in the woods, being still frightened of Caro. (Everyone else but Pari, that is, who went fearlessly to sleep in his hut. He told me so himself.) The next morning Caro said, "I'm *katu* now, I'm good, don't be afraid of me," and when he arrived at Canindé he was as courteous and gentle as João himself.

This was the first act in the drama. The second took place a few days later, and the cause of the quarrel was again a woman. Caro's dead son had left a widow: Siri, a girl so trim and plump and flirtatious that whenever she passed I felt like giving her a pinch. She had much the same effect on other people, and knew it, for she would waggle her hips and give warm, inviting glances as she walked by. Caro wanted her to come back with him and marry his younger son, the same son he had tried to marry off to Toí's sister. Siri didn't want to go with him, she didn't even want to leave Canindé, for she had got used to the life there, got used to eating fish and small cakes made from maize, to drinking coffee with plenty of sugar, and to spending the day in idleness. Not for her the work of a wife, digging up manioc tubers and making flour! So, when the day came for Caro to return home she came over to the big wooden house where João lived and sat down in the kitchen, saying she wouldn't go. Caro said that she must, and went down to the river with his men and loaded his things in a canoe to take him to the other side. Siri refused to budge. Then the fun started. Visiting Canindé that day was Anakã, the wife of Ingarassu. (She was no relation to Anakãpuku; Anakã is the name for a certain kind of parrot, and the suffix puku means long; Indian men and women often have the same names.) Anakã was not bloody-minded, like her husband, but she was a bit bossy; and when she saw Siri sitting there she began chiding. "Go, woman, go! You're not wanted here, go and get a man to live with, work, make flour!"

Siri sat tight. Anakã snatched her hammock from her hands—for Siri had been forced to pack her things up—raced down to the river with it, and gave it to Caro. Siri didn't move. Then two new actors came on to the stage, two young men who had come with me from Pari's village. One was Shiã, the goggle-eyed son of Anakãpuku, and the other his friend and rival in amorous matters, Kaitã, Pari's stepson. (His real father was Raimundo Caetano—the Tembé Indian who had brought about the pacification of the Urubus—as his name indicates; as well as his name, he got from his father considerable courage and character.) Shiã began to push Siri, saying, "Get a move on, Siri!" This was a mistake on his part, for Kaitã was Siri's lover—only the evening before he and Shiã had been fighting for her, and he had won. Now he came to Siri's rescue, and hit Shiã with considerable force.

Meanwhile Anakã had returned from the river, and seeing the two boys fighting together, sidled round them and caught Siri by the hair; she then dragged her out of the house, along the verandah, down the steps and fairly hurled her on to the ground. "Go, woman, go, go!" she shouted, still tugging at Siri's hair. Siri, sobbing, caught hold of Anakã's hands and said she wouldn't. Anakã lost her temper, picked up a large stick and whacked Siri violently on her back, still shouting "GO GO GO!" Siri picked up a pathetically small twig and gave Anakã two little taps with it. This surprised Anakã so much that she immediately stopped; besides, the rest of us were now all shouting at her and telling her to behave herself. Siri made off towards the jungle, weeping sorely—but more, I think, from frustration and rage than from pain—and didn't appear till late that evening, when she was found sitting on the kitchen table, her back turned to the world, whimpering slightly.

"Siri sa!" I said, to cheer her up. I don't exactly know what *sa* means, but it is some form of flirtatious endearment, and Siri understood it well enough. "Saé puku sa!" she replied, using my nickname—it means Long Man—and grinned weakly. Next morning she was showing her bruises with pride—five large blue weals.

The result of this affair was that there were three angry men. Anakãpuku was angry with Kaitã for having beaten up his son Shiã; Caro was angry first with Siri, because she hadn't come with him, and then with Toí, who took her back to his own village—as he had a right to do, since he was her half-brother—and Toí was angry with Anakã who had so thrashed Siri. The last I heard of the matter was that Siri had taken up with Tero. Tero had come to Canindé with Toí, to bring her home, and became so infatuated with her that, so it is said, he bundled her off into the bushes half an hour after leaving Canindé, and made love to her. But making love came easily to Siri, for I know that at least five men had enjoyed her favours inside a week before she left.

Ah, these quarrels! I think the Indians like them, for they bring a little excitement into their lives; in any case they are unavoidable, expressing as they do the combined strength and weakness of Indian morality. It is difficult to live with a hard man, who continually stands on his rights; and the hard man finds it difficult to live in a world where his rights are always being questioned. If only things didn't change! one can imagine him saying, if only there were no women! Then the world would be tidy, and everyone would know where they stood.

CHAPTER TEN

Measles and Death

SOCIAL order would be easy to maintain if it were not for sex and death. Sex among Indians always tends to become licence, and to undo society by ignoring its laws; death quite simply makes a hole in the social fabric which somehow has to be filled. They are not only fruitful causes of quarrels, as the story of Caro makes plain, but the two main doors into the supernatural world. But the supernatural is also dangerous, and to protect themselves from a fatal involvement with it the Indians surround both sex and death with a number of taboos.

Death is an especially fearsome matter, not only because it is so final, but because it separates a soul from its body. The soul then becomes a ghost, and should it decide to be troublesome (as ghosts will) there are only two ways of constraining it, both of them fairly complicated. One is to find the ghost a new body, in which it can be dealt with in a more or less straightforward manner: the other is to control the ghost ritually. Looking at this problem from the outside, one might say that the Indians—like all primitives—find it difficult to make the figures of their imagination correspond with things in the real world, and that they get round this either by objectifying these figures or by acting upon them internally by the process known as magic. If the Indians do neither of these things, then they are liable to be haunted: a part of their experience becomes detached from the rest, and tries to have itself recognised by fair means or foul.

The process of giving a ghost a new body takes several forms. One is through reincarnation: a dead man gets born again as his own grandchild, or great-grandchild, who is called by his own name. Many Indian tribes believe strongly in reincarnation: the Urubus, however, hardly concern themselves with the matter, though they like to name their children after noteworthy ancestors. Another way of giving a ghost a new body is best seen during a ritual execution, when an enemy prisoner is turned into a kind of Maír: it is described in Chapter 19. But the commonest way is to give the ghost, not a new body, but its old one: it is made to re-inhabit the corpse it has just left.

A ghost, naturally enough, doesn't like this. It is angry at being rejected from the world of the living, and jealous of those who survive it: and in this rage it becomes, as it were, peculiarly irresponsible and much to be feared, for the survivors are anyhow in such a state of grief and terror that the supernatural world can easily overwhelm them. I myself never witnessed a death, though I heard many accounts of death; and I saw something of the fear death inspires when news of a measles epidemic reached Pari's village.

One morning, at about ten o'clock, I was busy at my typewriter, writing up the events of the previous day. Chico had just finished washing his clothes, and was laying them out to dry on tree-stumps; Toí was sitting in our hut, for company's sake, making a strainer out of wicker. Suddenly he broke off from his work and pointed towards the jungle with his knife.

"Do you hear someone?" he asked.

We all listened. I didn't hear anything, though Chico thought he heard someone singing; then a dog barked in the old village, and everyone said, "That's right, listen, the dogs!" Then more noises—was it Shiã, perhaps, or Pari's stepson Kaitã? Then someone sang by our water-hole and halloo'd, and up came Tero.

"Did you hear anything down there?" I shouted. At

once there were shouts from the old village. "Go and see," I told Tero. Tero leapt out of the hammock he'd sat down in, swaggered ten paces down the path and yelled. Someone yelled back. More yelling, which I didn't understand. Toí's wife Terepik suddenly got to her feet and yelled too. "What is it?" I asked Chico.

"Dead," said Pari immediately. "Teré's dead."

"Who's she?"

"Shapi's daughter, she lives there," and pointed with a reaching forefinger.

Terepik was still shouting; Paijé joined in, and the noise was enormous. The men were all silent. "I don't want it! I don't want it!" was all I could hear from Terepik. Apparently it was the measles: and the bearers of the news were to stay away, she didn't want them anywhere near. It was the measles, and Teré was dead from it. "I don't want it!" she went on yelling.

Shiã, who'd been yelling to us from the old village, then came up and enlarged a little. It hadn't been Teré at all but Koso's wife—"You know Koso's wife?" Pari asked me.

"Ashira?" I asked, dismayed. I'd met her the year before: such a capable and pleasant woman.

"Ashira," he repeated with satisfaction—I think he had forgotten her name for the moment.

"Ashira!" repeated Paijé and Terepik, who went on talking shrilly to Shiã.

"Who's there," I interrupted, "who brought the news?"

"Kuata! He wants a medicine from you!"

I picked up my medicine box and went off down the path, followed by Paijé and three young girls, who bounded along with shrill cries. At the edge of the old clearing I saw Anakãpuku with his family, talking to Kuata and three other men who were squatting on the other side of the stream, some twenty yards away.

"Are you all right?" I shouted to Kuata as I came up.

"All right!" he said, getting up.

"Ashira dead?"

"Dead! She died in my village, in the place you drank cauin, remember? Now she's buried." Kuata, a shy but friendly young man, smiled pleasantly at me.

I asked if anyone else was ill. Kuata held up his fourth and little fingers, and said two people had fever: they had caught the measles at the same time as Ashira, but luckily recovered. I gave him some pills, and told him how they should be taken; and then, as Anakãpuku began to talk to him very loudly, so that neither Chico nor I could understand what he was saying, I broke in and said I was going.

"Aha! I go!" I said, politely announcing my intention.

"Nde! You!" Kuata replied courteously. "Come and visit us, we're nearby."

Paijé lingered behind for a moment for a chat, but overtook me before I had got back to Pari's. As soon as she entered the clearing, she began to shout, telling Terepik what she had heard. When she had finished, Terepik repeated the story, with equal zest, to her husband Toí, and to Pari, both of whom had been listening all along; and then Pari told it to Chico and me, in case we hadn't heard.

It was really a simple story. Ashira had caught the measles, she was the last one to get it, after everyone else had recovered: she got worse and worse till she was so ill that the others said, "Let's leave her, we'll go away!" So they left her early in the morning. Ashira woke up soon after and called out, and no one answered; called out again, "Koso!" but no one answered; called out to Kuata, no one answered. So she got out of her hammock, weak as she was, untied it, and with it hanging over her arm stumbled through the village saying, "Where are they? where have they gone? where are you? where are you?"

Paijé couldn't resist this part of the story, and interrupted Pari; she went to the edge of the hut, where Pari had a very small hammock in which his brindled hound would sleep, untied it and put it over her arm; and then, making a long face and walking with a staggering, floppy kind of walk, she made her way out of the hut coming to a halt wherever a tree-trunk lay handy on the ground—"Where are you, where are you?" she said, imitating Ashira, whimpering in a voice quite without expression.

That was how Ashira had gone along the path to the next village, and halfway along she'd fallen down and died. She was found there a few hours later, carried back and two days later was buried in a deserted garden near the village. Chico giggled slightly as he listened, that sad giggle which is a protest against having to take things seriously, but Paijé took no notice of him. "They all left her," she said, "they left her alone to die. What a death, I feel sad for her."

"I'm sad for her too," said Terepik. "All alone."

Paijé was very disturbed about the measles. She thought someone should go and tell Nusir, a member of Kuata's village who had gone on a visit elsewhere, about Ashira's death; and she had told Kuata to kill all Nusir's hens, in case he went back for them and caught the disease. "You mustn't go there either," she told me, "you'll only bring the sickness back with you and people will die."

Pari agreed. "Yes," he said, "Francis, don't go! You'll catch measles and give it to my wife."

"To my daughter, to everyone!" Paijé added. For she remembered another measles epidemic some years before that had killed many people. The survivors eventually ate dog's dung on Chic' Ourive's advice, as a preventive measure. "I'm going to eat dog's dung now," she cried, "so that I won't catch the measles."

"So am I," said Terepik, "we'll all eat it."

"Ugh!" I said.

"It's not what you think," said Chico, "you make a drink of it; you burn the dung till it's all ashes, then you mix it with water and put in some herb to make it taste nice, and pronto! you drink it!"

"When my husband died . . ." Paijé said, and began to reminisce. Her husband had been the village chief; he was called Pihun, Black, because of the colour of his skin. As he lay dying, the men in the village went off into the jungle, one by one, to be out of the way; but the women stayed behind, crowding into Pihun's hut. When he died, they let loose a great wail, and the men came back slowly. Yawaruhu (brother to Antonio-hu), who was one of the first to appear, shouted to Paijé that he wanted Pihun's feather collar: he shouted because he was standing twenty yards away from the hut Pihun had died in, not daring to come any closer. Paijé said no, she wouldn't give it him. Yawaruhu said she must; and then Anakãpuku returned from the jungle, where he'd been hunting. Yawaruhu put an arrow to his bow-string when he saw him arrive, and told him not to come any closer: but Anakãpuku came up in a rage and cut the iron head off the arrow with a machete, striking Yawaruhu on the back with the flat of its blade for good measure. Then he went up to the hut to see Pihun lying dead in his hammock. While he was there Yawaruhu started threatening Paijé again—"Give me the collar or I'll beat you!" and she gave it to him. When Anakãpuku came out, Yawaruhu was strutting about the village with it round his neck, saying, "You and your dead man, go and eat him!" Toí got the rest of Pihun's feather ornaments.

Then there was the time Koso's brother died, Paijé continued: they were so frightened of him dying that they buried him before he was dead. Paijé stood up to show what the man had looked like: she made a kind of Lazarus figure, her arms at her sides, with staring eyes and a tongue popping out between puffed-out lips, as still as a hanged man. They shovelled earth into the grave when he was still living, she said, and they could hear him gasp there inside.

Tero had been treasuring something to himself all this time, and at last couldn't keep it in any more: he tumbled down on to his knees, the top of his head pushed into the dust between his elbows, his buttocks riding high—"That's how Ashira died!" he said, and waggled his bottom; then got up, smirked and said, "If we'd come along the path behind her we'd have . . ." and he gestured obscenely—"with her dead!"

"With her dead!" echoed Pari, in astonishment. Paijé looked on, nodded and suddenly sat down on a log.

"That's why I don't marry again," she said to me inconsequentially, "because if my husband was ill, I'd have to stay near him while he died, and I'm afraid."

Immediately a man is dead, and the women have raised their great wail, the mourners cut their hair in a tonsure—or rather, used to, for few do nowadays—they burn some arrow cane and with the charcoal paint lines over their foreheads and down each cheek. Then the corpse is tied firmly into its hammock with lianas, and is carried to a corner of a capoeira, where the grave is dug. The grave is a very large one, some six feet long, four feet wide, and five feet deep. Two strong saplings are cut from the forest and planted firmly inside the grave, at the head and foot, so that the hammock with the dead man inside it can be slung between them. They put most of the dead man's possessions down in the grave with him, his favourite knife, perhaps, his bows and arrows, his club (in the days when they had clubs) and, apparently, his feather ornaments, which the corpse is made to wear. A whole basket of manioc flour would go down too, a gourd full of water, and the dead man's favourite dog, still alive, so that he could go on hunting in the sky.

A foot or so below the surface, a platform is built, supported by lengths of saplings pushed into the sides of the grave, and covered with large leaves. On top of this the earth is shovelled back, making a decent-sized mound; above the mound a miniature hut is built, thatched with palm fronds, with a small fire lit underneath. Many other South American tribes actually bury their dead in the houses they once lived in, which are then abandoned. The Urubu also abandon their houses, and if the village chief has died the whole village moves to another place, and the old huts are burnt down.

After finishing the grave, the mourners cut a great deal of brushwood, from thorny trees if possible, and pile it in a large heap on the path between the capoeira and the village, so that the ghost cannot pass. They do the same at the village, where every path may be barricaded by branches for some days. The mourners then go down to the river where they build a kind of stage over the water, apparently out of the living trees which grow on the bank; it may be thirty feet high. They carry water in large gourd bowls to the stage and there wash themselves, wash the charcoal from their faces, and let the dirty water fall back into the river.

These rites accomplish several things. First, the corpse is installed in its roomy grave and the hut built over its head, as if to show that here is its last home. The mourners blacken their faces with charcoal, to protect them from the ghost, partly by disguising themselves, and partly by the magic properties of charcoal which keep ghosts at bay. But the ghost cannot really be deceived by the disguise, for it tries to follow the mourners home to the village and is only stopped by the fences of thorny bushes which are erected on all the paths. The ghost, rather than wander aimlessly through the jungle, then turns for the only place it can call home, the grave. There it stays, 'waiting by the bones', as the Indians say. (This ghost, one must add, does not represent the entire soul. It is the bad soul, a kind of emanation from the corpse representing the decay of matter. The good soul represents what we know as 'soul', and it rises into the sky into the afterworld.) Having laid the ghost as well as they are able, the mourners wash the pollution of death from themselves so that they can start life afresh.

But the dead, even though one of their souls is supposed to go to the afterworld, and the other to settle down at the grave, still go on living, in a manner of speaking, for some time. They even quarrel among themselves, and a dead chief will feel extremely offended if he is buried close to a young girl or even a young man. A dead man is a jealous man and there is little he can do but live in the past. If he has been a great warrior, if he has walked through the jungle a great deal and seen the world, his experience sticks to him and gives him power; a power which only really dies when the fame of the dead man is forgotten. So he is buried by himself, in the most isolated corner that can be found for him, lest his injured and revengeful ghost come back to haunt his relatives. Children, on the other hand, can be buried close to the village, where they can feel comforted; they have done nothing in their short lives to make people afraid of them.

The last thing to be done for the dead is to revenge their deaths by killing an enemy. This propitiates in some measure their anger at being dead; their spleen finds an object and is thus diverted from their relatives. The custom is now going out of fashion and it is difficult to know just how important it may have been to the Urubu. But one knows how it worked among the Tupinamba, whose other customs are so like those of the Urubus. There the revenge took the form of a ritual execution, the prisoner being killed in honour of the dead man: and till this was done, the widow did not dare to remarry, for not only would the dead man come to haunt her, but her own children would be angry with her, and despise her for bringing such dishonour upon them. I don't think the Urubus had such positive ideas. A death in the family makes them feel very sad to start with, so sad that they won't eat anything for days: then they get angry, like Caro, and try to fill the hole in their lives by making a hole in someone else's. Death breeds death, and revenge revenge: but this is perhaps just as much part of the social order as anything else.

This small village in the jungle has about thirty inhabitants. The feast house is on the right, behind the trees; in the foreground is a corner of the manioc plantation.

At the rapids, the boat is poled up between the submerged rocks.

Chico, the interpreter, skins a howler monkey while a young boy does a derisive dance of triumph in the background.

In making manioc flour, the mash of tubers, after having the poisonous juice squeezed out, is put through a sieve, then roasted dry.

A man squats beneath his hammock to drink shibé.

Tero found that he hardly came up to my shoulder, which

Everyday male dress: feather through lower lip, collar of

Kuata sews a feather headdress together.

This woman holds open the beak of her tame parrot in order to feed it directly from her own mouth.

Men wear their best clothes to a feast, and carry bows and iron-headed arrows. Antonio-hu in centre; Wirakangupik at his left.

Saracaca (*second from left*) leads a line of women and young girls in a dance as they sing the duck song.

In the name-giving ceremony, the godfather, wearing feather ornaments, takes the child in his arms to begin the ritual dance.

The decorations on this young wife's comb, like her collar, are made of blue, red, and orange feathers. The black stripes are painted with genipapo juice.

CHAPTER ELEVEN

Sex and Saracaca

INDIAN women may behave badly, but it is not often that they create such trouble for everyone as Siri did, in the last chapter but one. Sex is not an event, like death, but an activity that goes on all the time; and its effects, though often insidious, are also diffuse.

Indians are such gossips that I don't suppose an affair can take place without all the women in the village knowing about it the same day; the news presumably gets round to the men later, but like men everywhere they are more interested in the future than in the past. At any rate, within a week of my arrival at Pari's village I had learnt about most of the young men's sexual adventures, not from them but from the women, especially from Paijé who had an enormous fund of such stories. She would come early in the morning to bring us our shibé, and to drain my coffee-pot that stood in the embers of the fire; then, if I gave her half a chance, she would sit down in my hammock and start chatting. I always wished she wouldn't sit in my hammock, for she would wrap herself up in it, or lie on her side with her head and right arm hanging out, while her dirty feet made a wonderful mess inside. Dirty feet, dirty hammock, practically the first thing one learns for oneself in a village.

So Paijé lay in my hammock, or sat astraddle in it, gabbling about sex. Stories of Kaitã and Shiã mostly, all about sex; sometimes about Tero, but all about sex, and all of one pattern. A man follows a woman, jumps on her, suruks her. A man waits in hiding till a woman comes along to get some water from the stream, jumps on her, suruks her. Like Kaitã perhaps, who followed an old idiot woman called Narĩ, old enough to be his grandmother: she went to the stream, where Kaitã helped her to fill her gourd pot with water. Then he emptied it when she wasn't looking, and set it afloat downstream. 'O Kaitã, Kaitã, where's my gourd pot?' And then he jumped on her. Like Kaitã and Shiã; someone in their village had killed a deer and cooked it, and the two boys had gone down the path and met a woman. She asked them, 'Is the deer cooked yet?' 'Cooked, it's all raw,' they replied, to stop her going to the village; and then they jumped on her and suruked her. Like Arishid, an eleven-year-old boy, who followed a married woman down the path, jumped on her, suruked her.

I listened to these stories in silence, nodding my head sagely from time to time. Not so Pari. "Young men are like dogs," he burst out, "they suruk women before they're men. When I was young, I didn't suruk a woman till I was so high——" marking a place in the air: till he was twenty or so, I suppose—"I was afraid of women! And now, when I go visiting other villages, I don't run after the women there, I sit down in a hammock and talk. People said that when I was with you last year I suruked lots of women. I didn't. It's a lie. Aren't I a chief? But the young men are like dogs. Tero, and Shiã, and Arishid, and Kaitã, they suruk when they're still children."

Paijé heard him out and then began again, recounting further impersonal encounters between young men and old women; recounting them with a kind of envious glee I found rather unpleasant. I wondered if anyone had jumped on her recently, but I didn't like to ask. She was getting a bit stout, and her face had a greedy expression on it—but this, from what she said, would hardly deter any young man from jumping on her if he wanted to. Presumably she did not meet young men often enough; besides, she was a widow, and had no one to comfort her at night. A few days later, however, she quarrelled loudly with her daughter Terepik over a man. It seemed that Paijé had been making flour one afternoon, when Tahi (the thirty-year-old son of Toí's miserable sister) came back from the jungle, where he'd been cutting down palms for the walls of his hut. He asked her for some flour to make into a shibé; then he asked for something else, and got it in an empty hut nearby.

"Tahi suruked you!" Terepik shouted accusingly, later that evening.

"He didn't," replied Paijé angrily. "You say that because you want Tahi to suruk you, and he doesn't. You're jealous! Toí isn't your husband any more, Tahi's your husband." Terepik was so annoyed that she didn't speak to her mother for days.

It was Mbeiju who told me this, Paijé's other daughter. Mbeiju thought it all a great joke, and giggled so much as she was talking to me that I could hardly understand her. Pari, on the other hand, put on a serious expression and said that Tahi would surely turn into a cururu toad after his death, instead of going to the afterworld in the sky, for such a gross act of fornication. Fornicators and adulterers, he said, all turned into cururu toads after their death, which I did not think unlikely, hearing the numbers of cururus that croaked every night round the village.

"Do you know about Muru?" he asked me suddenly. Muruambir, le feu Muru, was I suppose the most famous adulterer in the whole tribe, and for some reason everyone is quite positive he is now a cururu: perhaps he was too shameless to be thought of us anything else, for the woman he was always making love to was his *shuyan* or sister-in-law, whom he should have respected as fully as he did his mother. I thought it really a very sad story.

Muru, Pari said, had the largest penis that was ever seen on a man. It was so large that there was only one or two women who could take him, and one of these was his shuyan. Even so, she found making love to him a painful business, and in the end she couldn't bear it any more. One day Muru came pestering her when her husband had gone hunting. "No, no," she cried, appalled, and ran into the woods where she climbed a tree. Muru followed, but couldn't find her, and all afternoon he wandered about calling—"O shuyan! O shuyan! where are you? I want to sleep with you." She didn't answer. Two hours later her husband came home with an agouti he'd shot. "Wife," he cried, "come and cook this agouti for supper." When she didn't answer, he began shouting and went into the woods to look for her: and there she was up a tree. "What are you doing there?" he asked, so she told him. "Eh, Muru!" he said, and left it at that. But Muru, when he saw the two of them having supper together, he was sad; he got into his hammock and stayed there without eating, without talking, just thinking that there was now no woman in the world who would be kind to him.

I forgot to ask whether women also turn into cururu toads if they commit adultery. It would hardly be fair if they didn't, for young men could hardly succeed in jumping upon women if the women didn't let them, though this was something that Paijé preferred to pass over in silence. Besides, there are some women whose appetites are large as Muru's penis, and who do not wait for advances to be made to them. Such a woman was Saracaca, whom not even frequent beatings could restrain. I was able to see her at work, for she got a passion for Chico, my interpreter: and Chico, who became quite lovesick in return, told me how the affair progressed, for he could hardly stop talking about her.

Saracaca lived in Antonio-hu's village, and Chico first saw her on the day Antonio-hu came to visit us, that day Toí had all his beads stolen. She was being rather demure at the time, for her husband was with her; besides, she knew her place, for women keep silent on a visit when their menfolk engage their host in conversation. But she had not gone unnoticed, for I saw Chico eyeing her from time to time shyly but approvingly. She was worth eyeing. She still had something of a waist, her breasts were plump and firm, and the expression on her rounded face was attractive, if sometimes inclined to be at once heavy and sly. She took good notice of Chico, too, though she was standing some distance away; and the next time Antonio-hu paid me a visit I was not surprised that she came also. This time she had on her best blue skirt, ornamented round the waist with orange toucan feathers; her face was painted with red urucu, and her body, arms and legs with blue-black genipapo juice, in elegant stripes. On her back she had a leaf basket full of shelled bacuri fruit which she carried into my hut, brushing against Chico as she did so, and offered to me. Then she turned, again brushing past Chico, and went to sit on a log with some other girls under the eaves.

Antonio-hu and I had a pleasant talk about nothing in particular, and he left shortly after I had given him and the people who had come with him a few trifling presents, in earnest of better ones to come. Paijé immediately came over from her hut, sat in my hammock and complained—why was I always giving beads away to people who never did anything for me? There she was, bringing me shibés every day, and tapioca once or twice a week, and I'd only given her a few measly handfuls of beads, while pretty girls like Saracaca who brought me nothing but a few rotten bacuri fruit went off with bucketsful.

Paijé, of course, was jealous. I was always scrupulously fair in giving beads, in addition to slipping a small handful now and again to people, like Paijé, who had done me services; to Saracaca I had given no more than to anyone else. It was true, however, that Saracaca had more beads than other women, for she would extort them from her admirers, of which she had a number. 'Come and sleep with me!' they would ask; and when she had, she would finger the small ropes of beads round their wrists, and say—'Give me your beads.' How could they refuse? But Paijé never had such luck. So she began to tell me something of Saracaca's scandalous life.

At that time, Saracaca was living with three men. Her husband was a thin, serious fellow called Alishandre, whom I rather liked; her lovers were Mashĩ, a plump, shifty young man, and Antonio-hu. Mashĩ was the favourite; Saracaca insisted on cooking all his meals for him, even though this should have been the duty of the girl Mashĩ was about to marry. This girl was Antonio-hu's daughter, so that the situation was rather complicated; in addition, Antonio-hu was affianced to Alishandre's niece, and—the last straw—Mashĩ was Saracaca's step-brother.

Saracaca usually managed her affairs discreetly enough, so that her ménage prospered; but once in a while Alishandre would find her out, and beat her hard. Paijé, her eyes agleam, told me one such incident. One afternoon Saracaca was in her hut, spinning cotton, when Mashĩ came along, sat down and looked at her inquiringly. Mashĩ's mother was nearby, and she said—"O Mashĩ, make me some feather tassels for my new comb." Mashĩ, however, looked at Saracaca and said he was going for a bathe. Two minutes later, Saracaca put down her spindle and sauntered off to the stream as if she wanted to have a bathe, too; shortly afterwards Alishandre came in from hunting, got suspicious, and went to look for her. "Mashĩ! Mashĩ!" Mashĩ's mother called out fearfully, "here comes Alishandre!" But Mashĩ didn't hear anything, for he was rolling on the ground on top of Saracaca, and they were laughing he he hee! he he hee! When Alishandre heard this he lost his temper, ran up to them and gave Mashĩ a whack on the buttocks with the flat of his machete; and on Mashĩ running away he turned his attention to Saracaca who lay dishevelled on the ground, and beat her till she wept. "Now come," he said. When they got back home he pointed to the agouti he'd just shot and told her to skin it. "I won't," she said, "I won't, you've just beaten me. I won't do anything for you ever again." So he beat her once more.

"Alishandre must be sad to have a wife like that," I said. But Paijé said that he got all he deserved, being no better than his wife: his second nickname was Yawar, Dog, because he'd make love to any woman, any time, any number of times, anywhere.

Chico by now was getting really interested in Saracaca, and thought he might have a chance with her. He began to ask me when we were going to return Antonio-hu's visit, and if I wouldn't give him some beads after all (for he had refused an offer I'd made him before), and when I said I was going in a couple of days, he became very gay. Alas! the next morning Anakãpuku arrived from Canindé, bearing bad news. Chico's old girl friend had died: her father, a Tembé Indian living in Chico's village, had killed her by knocking her on the head with a stick. Anakãpuku roared with laughter as he told the news, and chaffed him brutally.

"Dead!" said Anakãpuka jovially, "and she was going to be your wife!"

"How do you suppose it happened?" I asked.

"Her father was jealous," said Chico, "they say he was jealous of her."

"But how, jealous? Jealous of his daughter?"

"Jealous," he repeated. Perhaps, he went on, her father had seen her going into the bushes with Emiliano, who was always after her, and might very well have beaten her when she came home. He wouldn't dare with Emiliano about, Emiliano was too strong.

"Dead," Chico sighed, "and all the money I wasted on

her . . ." He got into his hammock, and lay face down, looking at the dust below him, remaining in that position for an hour till I made him laugh by unintentionally killing my pet toucan. It had been given to me as a fledgling, and I'd got very fond of it: an absurd bird, still half naked, with its enormous beak and the tail it would cock up in an acute angle over its back when it went to sleep. At sundown, I fed it with some rice and bacuri, and some pieces of meat which it swallowed avidly; and as I put it up on its perch among the rafters, I somehow knocked it, and trying to save it knocked it even harder down on to the ground. Its crop was so full with food that it broke a blood-vessel; after gasping sadly for a minute, it expired.

It was now my turn to be melancholy. "You've killed your child!" said Anakãpuku, highly amused. I threw the corpse to Tero, who immediately plucked it and put it over the fire to grill. "Come and eat your child," Tero called to me when it was cooked: and so, inviting Chico and Anakãpuku and Pari to join me, I did. It was tender, but there wasn't much to go round.

Chico became a little less sad then, and Mbeiju further helped to distract him by telling him more stories about Saracaca: how, for instance, when Kaitã had been a young boy, Saracaca had said "Come and piss," and they'd gone together into the manioc plantation where she'd taken his hand, put it between her thighs, and pissed on it. "What are you doing?" Kaitã shouted wrathfully, took a stick and beat her till she fled. "Saracaca's crazy," said Mbeiju.

Pari warmly agreed. Once Saracaca tried to get him to suruk her, but he wouldn't, he was afraid; she came up to him and grabbed at his penis (here Pari inaptly demonstrated on Mbeiju), but he wouldn't let her have it. She then did the same to Kaitã, grabbed Kaitã's penis so hard that it hurt, and ran off with him in tow all the way to the stream. "It hurts, it hurts," shouted Kaitã as he ran, and when at last she let him go he didn't suruk her, no, he picked her up and threw her in the water.

Chico and I set off the next morning to visit Antonio-hu. Chico, when we reached the edge of the clearing, took off his travelling clothes and put on his best blue shorts and a clean shirt, washed his face and combed his hair. After this preparation he was very disappointed that Saracaca was nowhere about, that she had apparently gone with the men into the jungle to get tortoises and gather fruit. But Saracaca was in the plantation, and soon came back; when she saw Chico she retired into a hut to put on her best skirt, and to paint her face. Then she invited him to see the hammock she'd just woven for Alishandre, and soon they were sitting in it together holding hands.

"Ah," Chico told me afterwards, "there I was in the hammock, pinching her . . ."

"I saw you," I said.

"I thought you weren't looking! I thought there was only me and her in the world! And then Antonio-hu came back from hunting, so she went to the flour house, and I followed her; and I said, 'Ne ker akotar-in? Will you sleep with me?' Yes, she said. But, oh, defeat! oh, miserable one that I am! I had no courage. 'Come and get firewood,' she told me, but I didn't dare. Damn it," said Chico, who then turned himself over on his side and shook the hammock he lay in. "Damn it."

But Saracaca hadn't let matters lie there. After supper, Chico and I had gone over to talk to Antonio-hu in a small hut at one end of the village, a hut which had been walled in at one end to form a small room. We sat down in the other end, on a crude bench set against the wall, and Saracaca, I soon realised, was inside the room, doing something to distract Chico sadly from his job as interpreter. She was tickling him, poking her fingers through the gaps in the wall: poking him, poking him, till his side was all achy. Then she somehow got hold of his big toe, and pulled it till he thought it was coming off, and when she'd had enough of that she put her hand up his trouser leg and fondled his thigh.

"Chico!" I had to say several times, "pay attention, I can't understand Antonio-hu, he talks too fast." But Chico was in no state to interpret for me, for Saracaca was whispering to him all the time—"Why don't you make love to me," she whispered, "come and stay with us, don't go back to Pari's. Mbeiju's no good to suruk (for she was jealous of Mbeiju who saw so much of Chico), Mbeiju's thin, her carapua's much too small; but I have a carapua, a big one, I have a great big carapua, just try it!"

Chico went to bed that night in a daze. "Eh, senhor Francisco," he confided to me, "but that Saracaca, she's almost as bad as a Tembé girl."

"How are Tembé girls?" I asked.

"They come after you all day, they fondle-fondle you, they run after you in the woods and dig you in the ribs with their fingers, they *say* things to you: and unless you go quickly to their fathers and ask for them in marriage they call you a weakling. But call them whores and they'll hit you, hard, hard."

Back in Pari's village, Chico would sigh for Saracaca, remembering what she said, how she looked at him—oh, how she looked at him! He even thought of taking her away from her husband and settling with her in his own village. The Indians didn't think much of this—"Don't marry her," they told him one day, when he'd been a bit more lovesick than usual. "Don't marry her! Don't take her, she doesn't work: she doesn't like cooking tortoise, she doesn't like making flour, she'll give you her shit to eat. Don't take her!" This was a slander on Saracaca, for she was by no means lazy, and Chico ignored it.

"So what," he said. "These caboclos here (using the

derogatory name for Indians current among Brazilians) don't know how to treat a wife. If she doesn't work, I'll take a liana and beat her till she does."

"Perhaps she'll beat you back," I suggested. Chico snorted derisively. "Well," I continued, "you look out, she's cheated her husband often enough, she'll cheat you."

"Perhaps her husband doesn't treat her well," Chico said.

An Indian interrupted us: he leant forward and said, "Tero there, you see him, Tero? He suruked Saracaca and made a child in her, out came a thing without any face! like a devil! It had no eyes, no mouth, no genitals, nothing. Have you suruked her? If you have, and you marry another woman, she'll bear children without faces too. Look out!"

"How many children has she had?" I asked.

"Two, Tero was the father of one . . ."

"I wasn't," said Tero. "It was my brother."

"It wasn't, it was Tero . . ."

"It was my brother," said Tero in a small and feeble voice.

"Tero it was—and Saracaca gave birth to a devil, a howler monkey. She'd had another baby before that, and that was like a howler monkey too. Alishandre was the father."

"Oh," said Chico, "two husbands and each time a howler monkey baby."

Pari said something to me from the other side of the hut, but I couldn't hear: so he came over and took my head in his hands. "Ugly children," he said, "a man mustn't look at a woman who bears ugly children. Perhaps he can look just a little bit, but when she looks at him then he turns round and hides his face. If he looks into her eyes then all his children will be ugly too."

If this was meant to frighten Chico, it didn't succeed: he'd looked so long and so hard at Saracaca already that he felt one more look couldn't do any harm. And if it was a warning against adultery, to me it seemed slightly immoral and to go about things the wrong way. But Indians have different standards, and only trouble about morality if they themselves are wronged, or if it becomes involved with the spiritual process that makes a man hard: as the next chapters will show.

CHAPTER TWELVE

Puberty Rites

When there are no feasts or ceremonies in a village to give life a sense of occasion, and no visitors come to gossip, things often get very dull. The great recreation then, without a doubt, is sex; and young men, when they start feeling bored, go to the neighbouring villages to try their luck with the girls there. Still it is odd that people who enjoy so many sexual adventures should like talking about sex so much. Paijé was always on about it, and so, for that matter, was Antonio-hu. I went to stay in his village for some weeks, and one morning I was rudely woken by him sitting on top of me in my hammock. He put his face close to mine, made a bawdy gesture and asked if I was thinking about women. I grunted; I like Antonio-hu, but his breath smells. "Get up," he said, "get up and have a bathe! When we were young our grandfathers told us, 'Don't lie in your hammocks thinking of carapuas, go and bathe!' "

I got up and had a bathe. When I returned, there was a huge gourd bowl of tapioca waiting for me, sweetened with sugar. As I ate it, Antonio-hu insinuated horribly about the potency it would give me, rather like a fond mother urging her son to eat spinach. The lumpy squidgy bits, he said, were babies' eyes, and when I spat some out because it wasn't cooked, he laughed and said that now my children would never have eyes, it served me right.

After I'd finished the tapioca, he urged me to eat a great hunk of cold deer's liver, which I refused.

"Stop thinking of carapuas," he said.

"I'm not," I said peevishly.

"I know how it is! You haven't got a woman. When a man hasn't got a woman then he's sad, he doesn't do anything: even if he's hungry he sits about moping and does nothing. When he has a woman, and makes love to her, then he's all right, then he's happy, he gets up and does things."

"I've got a stiff neck," I said—the night had been very cold and misty. Then, as I spilt my coffee: "I'm kaú, I'm crazy!"

Antonio-hu knew how that was, too: some days he'd be so kaú that when he went hunting he forgot to leave the paths, he'd wander about hardly looking where he was going; or else, perhaps, he'd be sitting in his hut wearing a pair of old trousers, and would suddenly find his fly-buttons were open. "Now that's no good," he said, "that's no good at all, especially if the women see."

For all their big words, men are often curiously timid about women, and certainly are extremely ashamed if a woman should see them completely naked. One might think that an Indian, dressed up in the traditional way with a piece of string tying the end of his prepuce, could hardly be worried about looking naked: however, if a woman should see him without his string, he would nearly die of shame. A man's fig-leaf, in fact, is his prepuce which, when tied up, completely hides the glans penis, and it is when the glans is uncovered that he feels himself to be obscenely naked.* Men do not tie their prepuces when they wear trousers—there's no point, as they say themselves, since trousers hide more than a piece of string ever could—which is why Antonio-hu was disturbed when his flies came undone. But he felt ashamed to be naked even when there were no women around; bathing in my company, he would turn his back, slip off

* Men are so sensitive about this that they usually squat to urinate. Not so the women, interestingly enough, who, turning their backs, stand up bravely with lifted skirts.

his trousers, and sidle into the water with one hand over his crotch. When I walked about brazenly naked, he was quite put out.

I had myself tied up Indian-fashion several times, for the Indians were always wanting me to be a real man like themselves; and the women, of course, were always curious to know if a white man had the usual attributes underneath his trousers. It was perfectly true, the piece of string gives one a certain sense of security, and after a few moments I quite stopped feeling naked, at any rate till the mosquitoes and midges began to bite me; if anything, I felt a certain physical reserve, as though I was partly shut off from the world. So I could understand how a piece of string was as good as a pair of trousers, without understanding why men should be so ashamed of nakedness, even among themselves.

I could also understand why Paijé and Antonio-hu spoke about sex so much: one was a widow, the other a widower, neither had re-married, and both were feeling lonely. Still, other Indians had the subject on their minds almost as much, and would complain if they could not make love to their wives every day; and the women would complain also. I remember one night hearing a woman in the next hut talk in her sleep—"Husband," she said, in a moaning singsong, "I want your penis, I want your penis." But her husband wouldn't give it her, for she was ill, and to make love when you're ill can have fatal results.

It is indeed through such taboos that one can begin to understand the fundamental importance of sex. The Indians think of it (and, consequently, of the physiological functions connected with it), not merely as a pleasurable recreation but as a contagious spiritual force that, if it is misused, can have the same effect as a poison. When we suffered from diarrhoea, as we frequently did from eating too many fruit, Pari would at once conclude that some couple had made love under the bacuri trees and so blighted the fruit we had eaten—the fruit themselves, he said, were quite innocuous.

Fruit, the Indians think, are themselves the result of sexual activity—"but they haven't suruked yet, they're still small," as Antonio-hu would protest sadly when I asked him to find me some bananas—the result of a transformation that is really a continuous act, not a single one. Sex is what makes things grow: a magic. In one sense, of course, the real magic is not the sexual act itself but the transforming desire behind it which, so to speak, overflows during the act, and so infects the surrounding world. To make love to a woman under a fruit tree, therefore, is to charge the tree with an excess of growth-magic that spoils the growing fruit as effectively as hormone weed-killer kills a growing plant. The same kind of reasoning was behind Tero's complaint one day that some deer he ate wasn't good to eat, for he'd noticed that the couple in charge of the cooking had been making love at the time. This was a silly thing to do, for not only did their love-making infect the meat that was then being cooked (and cooking is of course a transformation just as much as the ripening of fruit), but the cooking and eating of deer-meat is subject to particularly stringent taboos, to break which may well bring a calamity upon the whole village.

Taboos, as I shall explain more fully in a later chapter, are always imposed whenever a process of transformation is taking place. When human beings are being transformed, the process is usually called a passage-rite, being the rite that lets people pass from one status to another —such as the rite that gives a new-born child a name and makes him a member of the tribe, the rites that turn youths into men, or girls into women, or dead men into inhabitants of the other world. All these rites are hedged in by taboos, as are the minor processes of transformation such as cooking, baking pots or brewing cauin beer; and the activities tabooed are themselves always parts of a process effecting a quite different transformation. A youth, undergoing the rite that makes him a man, may not make love; a man baking a pot may not eat or defecate; a person eating deer-meat may not sing. Every such process is at bottom a magical one, and magics don't mix.

For the Indians, perhaps the most dangerous of all these processes is that of menstruation, the regular and spontaneous manifestation of the creative power as blood. Blood is the very principle of life, as the Indians acknowledge every time they paint their faces red with urucu, in imitation of it; for that very reason, however, it is dangerous. No Indian will eat half-cooked meat, lest the blood that is still in it should poison him; and if he should tread in another man's blood he will turn yellow (the sure mark of spiritual sickness) and die. Menstrual blood, being that substance out of which a child is made, is much more dangerous. Pari had been affected by it, for he had slept with his first wife, not knowing she was menstruating at the time, and as a result he had become *panem* or unlucky in hunting—and certainly it was true that he seldom managed to kill anything larger than an agouti or a paca unless it had first been wounded by someone else. I too had been affected by menstrual blood once, or so Pari maintained. It happened one day that I had a terrible attack of diarrhoea, for no reason I could think of. At first Pari thought it was caused by the tapir meat we had been eating, for tapir meat is notoriously 'strong' and often gives people stomach-aches. Chico, joking, said it was a punishment for not letting him go to Antonio-hu's village and see Saracaca. At length Pari's eye fell on a large bunch of bananas a woman had brought me the day before. Her name was Matakiki: an extremely energetic woman, who could have run a Women's Institute, had there been such a thing in the jungle, with complete efficiency. She was also very honest, and never asked for a gift if she had nothing to give in exchange—the bananas, which she'd carried from her village two hours' march away, were a mute appeal to me that I should give her some beads, some needles and thread, and even a knife if I could afford it. As she carried the bananas, however, she tripped over a root and so brought on her flux. Now menstruous women may not work, nor prepare food others are going to eat, lest they contaminate what they touch—they even use the oldest of gourd bowls at their meals, so as not to spoil their good ones—so that Matakiki, by rights, should either have eaten the bananas herself, or thrown them away; certainly she should not have given them to me. Rather than waste them, however, she took a chance, and left with suitable presents. I had eaten at least two of them, and Pari thought there was no doubt that they, infected by Matakiki, had caused my diarrhoea.

Women therefore go into partial seclusion when their courses are upon them, sitting in their hammocks and eating special food. A girl's first menstruation, of course, is much more important, for not only is she then manifesting the creative power, but she herself is being transformed into a woman who can bear children. The Indians talked with wonder of a young girl in a neighbouring village who actually conceived before she had ever menstruated, and were quite at a loss to account for it—the event was plainly unnatural. It was also wrong of her, they said, to have made love at that age, for her vagina might have shrivelled up and become dry and useless, or she might have sickened, become thin and ill. This is what happened to Mbeiju, who had slept with several boys when she was still a girl, and who now was thin and plagued by dreadful headaches. But her headaches, though she did not realise it, were connected with the fact that she had missed her periods for some five or six months: from the time, she said, that she had eaten some pig shot by her uncle. It was probably the pig, rather than the uncle, that caused the trouble: she may well have been beginning her courses on that day, and pig is not a food she should have eaten. However, the Indians themselves were baffled by the situation, and several of them thought it had all come about because she had made love when too young.

Mbeiju had several ways of dealing with her headaches, two of which, curiously enough, have a direct bearing on sex and passage-rites in general. The first method she had was to make herself sick by swallowing a quart or so of warm water. Another way was to get herself treated by an amateur shaman or medicine man, who would fumigate her with tobacco smoke, suck out the pain and banish it with decisive gestures and further volleys of tobacco smoke; and the remaining methods worked on the principle that one pain drives another out. She would take some nettle-like plants and rub them over her forehead and behind her ears—sometimes Paijé, her mother, would force her to rub them over her entire body—and if that didn't work, she would get Pari to scratch her with the lower jaw of a small fish that had sharp teeth, till the blood flowed, and then to rub pimento into the wounds.

The Indians think highly of pain and make use of it during rituals in order to increase their spiritual strength: it is all part of the idea that the true Indian is hard-bodied and capable of enduring anything. The warrior who has killed an enemy in battle gets a companion to scratch his body, as Pari did to Mbeiju, though pimento is not rubbed into the wounds afterwards; the man who ritually executes a prisoner for a cannibal feast has the same thing done to him, and must also endure the bites and stings of ants and wasps. There is no doubt that these practices do in fact make Indians strong, as they declare; but, while one must take such explanations into account, it is as well not to trust them overmuch. The Indians' world is a perfectly reasonable one, and the mythological connections between cause and effect are quite plain; the only trouble is that the Indians themselves are seldom conscious of these connections, for instead of explaining things in terms of motive they do it in terms of purpose. The purpose of scratching one's body with a tooth is to make one hard; the motive that makes the Indians elaborate this practice, rather than another one, is something quite different; it is a kind of self-punishment. There is no doubt that Indians feel a certain guilt whenever they kill people, and also, of course, when they eat them; not the guilt of a bad conscience, that they have done wrong, but the guilt of having disturbed the supernatural world and bitten off more than they can chew. To kill or eat another human being is to commit an act of hubris, one might say, for which a just retribution is exacted; a retribution that is not so much a vengeance as a re-ordering of spiritual forces. The Indians at such times make use of rites in order to co-operate with these forces, and bring about the necessary changes; the pain which they inflict on themselves is not meant to forestall supernatural retribution but is in fact the ritual enactment of it.

When a girl menstruates for the first time, she also is made to endure a certain amount of pain. Here again the reason given for this practice by the Indians, that it makes her strong, is by no means the whole story—guilty feelings are somewhere involved. The fact is that Indian men are jealous of their wives for being able to bear children: they like to think, as their mythology states plainly enough, that the business of creation is a peculiarly masculine affair in which women have no part. True enough, it is women who actually bear children, but it is thought that they do not actually make them, in spite of their beliefs about menstrual blood: men make children, women only nurture them as the earth nurtures a growing plant. Indeed, the word for semen is the same as that for a child, and a man says "my child comes out" when he ejaculates. When a girl menstruates for the first time, therefore, she is repeating woman's age-old offence of contradicting man's pretensions, by showing that—mythology notwithstanding—it is she who makes children, and not him. This is her guilt, and the pain she undergoes during her initiation as a woman lets her atone for it: by this means the contradiction between the masculine world of mythology and the feminine world of physical reality is resolved.

The gap between these two worlds is very real: mythology speaks of heroes but never of heroines, and whenever it does mention women, which is not often, it denigrates them. At the beginning of the world, for instance, women neither conceived nor bore children, for Maír could make them by himself. He would ejaculate into an earthenware pot, cover it up, and wait till he heard a scratching noise come from inside the pot; then he would know that the child he had made was fully formed and ready to be taken out. This way of making children, however, being a magical one, was only effective if no one peeked into the pot while the child was being made, so he warned everyone not to interfere. One day, however, a woman passed by Maír's hut, where a great pot was standing, and hearing a scratching noise come from it she became curious, lifted the lid and looked inside. The child immediately died, and after a few days began to smell. When he discovered what had happened, Maír flew into a rage, picked the child up and threw it into the woman's belly. "That'll teach you to be inquisitive," he said; "now you'll have to bear children, and when you do it will hurt." True enough, she carried the child about inside her for nine months, and when it was born she cried with the pain, as all women still do.

Such stories are a good way for the men to get their

revenge on the women. But the true state of affairs is told in another myth, in which men, for once, come out very badly; and even though it acknowledges that children come out of women and not out of men, it still tries to hide that fact as much as possible. It is the story of Maír's rankuaĩ-ang, his ghost penis, that existed at the beginning of the world, just after Maír had created men and women by drawing them out of the wirapitang or red-wood tree. In those days the women did all the work: they cut wood, hunted, cleared the jungle, planted manioc and built the huts, while the men lay in their hammocks and slept. The men were like children then, for none of them had a penis, nor did Maír himself. This made the women angry, so Maír made a rankuaĩ-ang for them that lived under the ground. It was an enormous thing, like a great blue worm, quite half a mile long; when a woman wanted it she would call softly, "O rankuaĩ-ang, o my man, my husband," and then stamp gently on the ground. The rankuaĩ-ang would hear, thrust its way through the earth and raise its head just in front of the woman, so that she could squat on it and take her pleasure. When she was finished, the rankuaĩ-ang would disappear; and she would take a large pot, piss in it till it was quite full, and put it away for five days, when she would find a baby inside it.

One day, however, one of the men became tired of sleeping, got up and followed a woman who was going to roast manioc flour. She lit a fire under the great clay roasting plate, and while she waited for it to heat up she suddenly got a desire for the rankuaĩ-ang. She called gently, stamped the ground, and squatted on the rankuaĩ-ang's head as it emerged. The young man saw all this, and, when the woman had gone, thought he'd have a try at summoning the rankuaĩ-ang. Sure enough, when he stamped on the ground it thrust up its head—and he, not knowing what else to do, took a machete and decapitated it.

"Oh, disgraceful youth!" as Chico said, when he heard the story. Two days later the woman went back to the flour house to roast manioc, stamped the ground gently, but nothing happened. She stamped again, still nothing happened. Rankuaĩ-ang was dead. In a fury, the woman went to Maír and shouted at him, "Where's our husband got to? Where's our man?" Maír didn't know. The women were so angry then that they filled the river with rocks till the water disappeared, and left to make themselves a new village many miles away.

Maír became anxious. "Who killed the women's man?" he asked.

"I did," the youth replied.

"Well, go and get him."

So the young man went to the flour house, pulled the rankuaĩ-ang from its hole, found its head—it was like a snake's head, with eyes, mouth and a forked tongue—and brought it all back. Maír cut the body of the rankuaĩ-ang into lengths, shaped each piece carefully, and fastened them on to the men. When he'd done that, he sent for the women—"Come back," he said, "all the men have a rankuaĩ of their own now"—and the women, nothing loth, came back, and were promptly suruked. But one man was very sad, for his rankuaĩ came off inside the woman he was making love to, and she pulled it out and threw it away. "Eh Maír, my rankuaĩ's come off!" he complained. Maír put it back on, and tied it in place with the bark from the embira tree: that's why men have pubic hair. As for the women, they could no longer make children by pissing in pots, but had to keep them in their bellies and give birth to them.

This story can be read in two different ways. It is first a myth for an adolescent boy, who has never made love to a woman and is afraid that his penis will not be equal to the task, when he does; though at the same time he day-dreams that he can enjoy any woman he pleases by stealth, that his desires, in fact, are omnipotent. It is only when he grows up, and his pubic hair begins to grow, that he can enjoy women in his own right, and not vicariously by way of the rankuaĩ-ang; and it is when he realises this that the rankuaĩ-ang is killed, and turns into his own penis.

But the story can also be read as the world-wide myth of the Rule of Women. Other South American tribes speak of a time when men had only a subordinate rôle to play in rites and ceremonies: the women had charge of the cult objects, and by their use terrified the men into subservience. At length the men revolted, put the women to flight, took control of the cult with all its paraphernalia, and by so doing became the dominant sex. Nowadays, in such tribes, it is the women who must hide themselves whenever the men make the spirits talk by means of trumpets or bull-roarers, and who are forbidden to see the masked dances or learn the secrets of the cult. It is impossible to say whether this represents a real historical event or not, but this doesn't really matter: even if the men use the myth only as a means of self-justification, the fact remains that they see in women a power that continually threatens them. The Urubus fear women also, though they have no cult which must be kept secret from the women: and what they fear, and are jealous of, is the creative power women embody, together with the sexual attraction that can make a man forget his spiritual ideals. Hardness is not only a description of masculine virtue, but of the erect penis: and when it is put to work in the service of women, the hard becomes soft, and the man loses his virtue. It is always women, meddlesome and inquisitive, who are the cause of his downfall from the spiritual to the physical, and from his point of view it is perfectly right that mythology should be full of warnings against women and all they stand for.

All this goes some way to explain the difference between the puberty rites for boys and girls. Since there is no secret cult among the Urubus, there is no group initiation for boys: for them the process of becoming a man is a long one, which they achieve through taking part in raids and by killing an enemy prisoner in a public ritual. The time of puberty is only important for them because they then start getting embarrassed with the size of their penises, an embarrassment which the women make more acute by flirting with them and pretending they want to be seduced. This only ends when the boys are thought old enough to have their prepuces tied, and till that time they feel very ashamed of themselves, and try whenever possible to cover their genitals with one hand. Nowadays they either tie up their prepuces themselves, when they feel life has become unbearable, or their mothers do it for them—that is, if they haven't already taken to wearing trousers, as most of the boys near Canindé have done—but in the old days it was a matter for the tushau or war-chief of the village. The ceremony would take place during a cauin feast: the boys would stand before the tushau, who would ask them, "Do you want me to put on my maracaja-skin cap for you?"—that is, if they wanted to recognise him as their chief and so become his warriors. Then each of them would give him some arrows, as a token of their allegiance, and he would tie their prepuces as a token that they were his warriors, that they were men.

A boy was thought fit to undergo this ceremony when his pubic hair was well grown. This by itself is a matter of some embarrassment: among many South American Indian tribes, the men pluck out all their hair, on their faces as well as on their bodies, as soon as it appears, as though they feared to advertise their manhood too fully. The Urubus let their pubic hair grow, but at least one young man I knew felt so abashed at the responsibility it implied that he borrowed my razor and shaved it all off. Pubic hair is the very mark of adult man: and here, for once Nature comes to their aid and shows how childish and inferior women are, for they rarely grow hair anywhere on their bodies. Tero once had a great time teasing Mbeiju—"She's got hair on her carapua," he said, ruffling the hair on her head, "like this, thick, thick!" Mbeiju was properly incensed—"I haven't," she retorted angrily, "it's a lie, I've none there at all, no." She would have been very undesirable if she had: Anakãpuku once confessed to me that though he would like to sleep with a Brazilian woman he didn't dare: they had hair between their legs, and the very thought of it made him frightened.

If boys are embarrassed on growing up, girls are not. They are proud of their breasts, and proud, if a little scared, when their courses first come upon them. They have had to work throughout their childhood much harder than the boys: it is their job to fetch water, cut firewood, to help digging up the manioc and turning it into flour. They have done all the jobs women have to perform, without enjoying a woman's status: now, when they menstruate and are put into seclusion, they are women in their own right.

Girls' initiation takes many forms. Chico told me one evening how the Tembés do it. The girls are all secluded the moment they begin to menstruate, and are put on a diet of small fish. When the flux has stopped, the women shoo the men out of the village, make a great soup of sweet manioc which the girls drink in order to heat their vaginas, as they say, thus making them fit to conceive.

Mbeiju listened to this with fascination. She was eating tapioca at the time, for she had one of her headaches and thought that the tapir meat we were eating would only aggravate it. As she gulped down her tapioca* she told

* Tapioca comes from manioc, like flour; the tuber, however, isn't put into water to rot slightly but is grated finely while still fresh, and the juice which is squeezed from the mash is put to one side and left to settle. What

me what happened to her when she first menstruated. She went into seclusion like a Tembé girl, but was allowed to eat yashi, the white tortoise, as well as two kinds of fish, mandji, the slimy catfish, and aracu, a fish with many bones; and she could also drink watery shibés made from farinha seca, as the Brazilians call it, the roasted mash that is left after tapioca has been extracted. All day she had to lie in her hammock fully stretched, for her feet had to be kept clear of the ground, lest the magic that was in her should escape in a kind of short-circuit. "She lies like this!" said Pari, interrupting her: he got up from the tortoise carapace he'd been sitting on, stretched himself out in his hammock, caught the sides and with them covered his feet. Like that, he said, she lay all day; when the sun was overhead she was given tortoise to eat, when it set she was given a shibé.

Mbeïju decided that she too ought to demonstrate the position for me, and got into her hammock. "We lie like this," she said "and we mustn't go to sleep. If we did we'd always be in our hammocks during the day, we'd sleep, sleep! and not work at all."

"And then," said Pari, "she had all her hair cut off, all of it, right off"—he gathered up Mbeiju's hair in his hands, close to her scalp—"it's all shaved off."

"And ants," I said, "do they put ants on her?" I had heard about it before: a cotton band is woven, ants are inserted between the meshes, the band is fastened round the girl and not taken off till the ants have become too weak to bite her any more, or have squirmed loose and fallen off.

"Ants," Mbeiju said with emphasis, putting her hands below her breasts.

settles is tapioca, needing only to be dried and then boiled to be fit to eat. The tapioca that comes to England is roasted, to make it keep better; but the Indians say that this practice is dangerous, for the person who roasts it will get such a knot in his anus that he will never be able to defecate.

"Ants," said Pari, "here, here, here," and demonstrated on Mbeiju, putting his hands on her forehead, round her chest, round her waist.

"Here!" repeated Mbeiju, in a kind of triumph, caressing her body. "To make one strong!"

Tapi-ai-ĩ, the ants are called: black ants, about an inch long, with large and venomous jaws. Their bite is not as poisonous as that of the tocandira, which can immobilise a man for a day or more, but like the sting of a large and vindictive wasp. They do not bite just once, but many times, till their venom runs out; and the young girls are made to stand up and endure the multiple agony of several score of these ants biting at the same time, without being made partly insensible by drink as is the custom among some Indian tribes. A girl who has endured this initiation has indeed paid her debt to mythology and become strong, and the satisfaction with which Mbeiju told me about it showed the confidence it had given her.

After the affair of the ants, Mbeiju had gone to the flour house to prepare tapioca. When it was made, she took it round the village, giving some to everyone present; then in the afternoon she made flour and distributed that also, not keeping any for herself lest she become mean. Her seclusion was then at an end, and till her hair grew down to her shoulders the only thing forbidden her was making love. "Mbeiju here," said Pari, poking her in the ribs, "made love before her hair was grown, that's why she's so thin. Suruk!" He lay back in his hammock, flexed his knees and threw himself into a pantomime of a girl making love for the first time, wanting it more and more, till the hammock could hardly hold him. "But when her hair's grown again," he continued with hardly a break, "and when she's had her sip of cauin, then she can suruk all right." When the time came, Mbeiju dressed up in her best skirt, put on all her feather ornaments, had her hair trimmed neatly and her face painted;

then she took a drink of cauin from a small gourd bowl, and went to sleep.

"And then you married her," I said to Pari, as we both looked at Mbeiju.

"I didn't," he said with a curious vehemence, "I never married her." He was quite firm about it. He'd been married once, to his first wife who had died some years ago, but he hadn't married any of the other women he had lived with. In the old days it was different, for a man and a woman who lived together always got properly married, were husband and wife. Marriage is an important institution, for several reasons. It is the penultimate step leading to the very purpose of the sexual act, the making of a child, and the rite by which this is made possible not only binds the man and the woman together in a special relationship, but makes plain exactly what their duties are towards each other.

The preparations for the marriage are all done by the bride's mother. It is she, helped by her husband, who grates manioc for the cauin, and watches over it while it ferments; then they invite the neighbouring villages to drink it and witness the ceremony. The bride and groom would be sitting in their own huts, perhaps at opposite ends of the village, and had to stay there while everyone else was getting drunk, without having even a mouthful of cauin. But when the cauin was finished—if there was a lot, it mightn't be for two or three days—the chief would summon the couple to the feast house as the sun reached its height, and sit them down in the same hammock. The bride would sit on the left of the groom, her right arm round his neck and her hand on his head: his left arm would be round her neck, and his hand on her head. The chief would take a piece of red cloth, wrap it about their heads, and say—"Now you are married." In this way they were literally bound together, the red cloth being a piece of magic ensuring that the union would be fertile. Then the chief would harangue the groom. "Don't beat your wife," he would say, in a formal, impersonal voice, "don't scold her, be good to her. Get tortoises for her to eat, plant a lot of manioc. Make children." He would continue in this vein for some time, detailing all the things which made for a good marriage, and when he finished the couple would go to the groom's hut. There the bride had to bring him two shibés to drink, and the groom had to provide some meat to cook: so they were married. Then for ten nights, Pari said with a grin, they'd make love—all night, every night, without ever sleeping, just making love till the man was exhausted. Then, perhaps, they'd sleep a little.

Mbeiju, slightly scornful of her 'husband's' pretensions in the matter, for he was occasionally impotent, got out of her hammock and sat on the ground again, spooning up her dreadful tapioca—plain tapioca boiled with too much water, without salt, without sugar, without even a flavouring of grated palm nuts.

"Mbeiju eating boiled tapioca!" jeered Chico.

"In my father's time," Pari said, "we never called boiled tapioca by its name at night, or we'd get pimples all over our bodies."

I was about to dismiss this observation as irrelevant when I remembered the use tapioca is put to during a girl's puberty rites. Was tapioca, I wondered, a kind of vegetable semen, with magical properties? Mbeiju was also interested, though for other reasons, for she had called her mess of tapioca by name several times that night. She took the small paraffin lamp and examined herself; sure enough, she found a lot of pimples.

"I told you so," said Pari. "At night you should call it agouti *mingau*, the agouti's porridge." Mbeiju laughed, for she suddenly realised that I'd spoken of her tapioca by name that night also, and to make sure I would get pimples, in a private place if possible, she gave me some of her tapioca to eat. She was really quite fond of me.

I discovered later that tapioca does indeed have spermatic qualities. I asked Anakãpuku if eating it gave one an erection. "Who told you that?" he asked with interest.

"No one," I said.

"Capitã Pihun thought so," he remarked, "he was always eating tapioca and he said it made his rankuaĩ *hantan*, hard as anything! Paijé knows. She was Pihun's wife, she knows." Mbeiju was the result of one of Pihun's gorgings on tapioca, for though his skin was black, as his name implies, Mbeiju has a very light skin, almost as pale as tapioca: the tapioca, therefore, had had a double action. So this makes it plain why a girl during her seclusion can only eat flour from which the tapioca has been extracted, and why she has to make tapioca afterwards and distribute it round the village. Tapioca is a magic substance that increases a man's potency, so that a girl, herself potent with menstrual blood, may not eat it lest the one creative process should interfere with the other; but she can make it, and give it to everyone in the village, as the sign that she is now fertile and as capable of making a child as tapioca is. Through her, the whole village is given new life.

As for calling tapioca by the name of agouti's porridge during the night, that seems to be a roundabout way of saying that man is a spiritual being, like the sun itself, during the day, and a sexual being when the sun sets. Supernatural forces must never be named openly when their power is at its height, just as you must never use a man's name when talking to him: it is a source of embarrassment, and puts him unwillingly in your power. Instead you just use his nickname, a euphemism for his most characteristic trait. The agouti is a notoriously randy animal, and the tapioca spirit can have no doubt it is being graciously referred to when you speak about the agouti's porridge.

CHAPTER THIRTEEN

The Moon's Vomit

INDIANS believe, like most people, in the existence of a Golden Age. Sometimes they think of it as being in the past, sometimes in the future, and sometimes in the other world, and they often describe it in contradictory terms. The Golden Age represents an ideal state, in which life goes on without effort, and this ideal state can refer either to a perfect working of the social order or to a complete absence of order. The Urubus, whose lives are so often disturbed by quarrels, wars and irruptions from the supernatural world, have utopian desires after order of which Maír, their culture-hero, was the inventor and is now the exemplar. The visitor to Maír's legendary village will find the best of everything lavishly supplied: game and manioc flour in abundance, and huge pots of cauin which can never be emptied, however much is taken from them. Maír sits in his hut all day, talking and eating, and when night falls and the moon rises he gets out his drum and begins drumming, or leads his villagers in a dance. The curious thing that immediately strikes one, however, is that though his village is full of children, none of the men have penises—just as in the story of the rankuaĩ-ang—and there is considerable doubt whether Maír himself has one. Some Indians, indeed, say that his penis is the largest in the world, having been made out of the head of the rankuaĩ-ang, but others say that it as small as a baby's little finger. It seems then that Maír leads a double life: he is a sexual power when he wanders through the world, but when he chiefs it in his village and has to be responsible for law and order, this gross attribute of his manhood vanishes.

If sex is always undermining the social order, it is not to be wondered at that Maír safeguards himself by abolishing it: for his order is based on distinctions of place and power, distinctions which sex can bring to nothing with the greatest ease. To make human life at all possible, however, a compromise must be reached between the two forces represented by Maír and sex, between the states of ideal order and ideal lack of order. This compromise is most easily seen in the kinship system, where sex is made to work in the service of order, through marriage; while order protects itself against sex through incest taboos of two kinds, the first forbidding all sexual intercourse between some relatives, the second only forbidding marriage between others.

The Urubu kinship system is simple enough, and a rough idea of how it works can be gained by looking at its terminology. A man calls his father and his father's brother by one term, *father*, his mother and his mother's sister by another, *mother*: their children are therefore his *brothers* and *sisters*. His father's sister is called *aunt*, his mother's brother *uncle*, and their children *cousins*. Since the preferred marriage among the Urubus is with these cousins, the term *cousin* also does duty for brother-in-law and sister-in-law; for the same reasons, parents-in-law are called *uncle* and *aunt*.

These fundamental relationships within the family are kept intact by means of an incest bar. A man may not sleep with his grandmother, his mother and the other women he calls *mother*, or with his sisters. There is no human punishment involved if he does: retribution, if it has to come, is always supernatural. I only heard of one man who committed incest, and that was a youth who slept with his grandmother: he became paralysed soon after, and died within the year. The incest bar, theoretically, should also apply to all the women a man calls sister as well as his true sisters. These *sisters*, however, who are really cousins, are forbidden a man not so much in sex as in marriage. The difference between them and the cousins a man can marry demonstrate the fact that the incest bar has really nothing to do with genetical inbreeding. It has, instead, two quite different functions. One of them certainly has to do with inbreeding, but it is the inbreeding of the social order and not of the physical one that is involved: by forbidding marriage between a man and the cousins he calls *sister*, the family group is made to contract alliances outside itself, and stops becoming isolated. Isolation is one of the great social dangers in the jungle, and a tribe (such as the Jívaros of Ecuador) in which family groups become separated and suspicious of each other soon becomes fragmented, and blood-feuds between villages are accepted as the normal way of life.

The other function of the incest bar is, in very general terms, to give society a sense of direction. Every system has its own internal laws which must be obeyed if the system is to work; the incest bar is the basic law of the kinship system, to infringe which brings a real death upon the community, since everything that its order implies is then denied. For an Indian only knows how to behave because he is related to other people in a definite manner, with definite obligations; indeed everyone in the tribe is classed as a relation of some sort, and even strangers, such as myself, are called *uncle* or *grandfather*, in order to give them an acceptable place in society and let people know how they should behave to them.

The crucial moment for society, which is the real test of the kinship system, is marriage: when two people, who are already related by blood, create a new relationship between each other by law. Every kinship system has its own particular solution for the problem of marriage, the solution being an expression of the incest bar; for the

Urubus the preferred marriage is that between cousins. Now, a single man is always slightly embarrassed when he meets one of his cousins—should he treat her like a member of the family, like a potential in-law, or like a potential wife? An even more complicated relationship is that between a man and his niece who, in the old days, he could also marry. Any woman would find it confusing to be a man's niece and his wife at the same time, and the confusion is also felt by her brother, who finds himself both nephew and brother-in-law. When Indians are confronted by this difficulty, they have recourse to a special kind of formal behaviour, such as the set ejaculations mentioned in Chapter 5; by this means they create a new type of relationship which allows close relatives, who might otherwise be embarrassed, to live side by side while keeping at a safe psychological distance.

What one calls the kinship system is of course an abstraction. For one thing, the marriage scheme which it implies represents an ideal state of affairs rather than an actual one: nowadays, few Indians marry their cousins. There seems to be a patriarchal tendency among the Urubus which makes them dislike such close marriages; instead the most frequent type of marriage is that between second cousins—that is, the children of cousins. In the same way avuncular marriage—common among polygamous societies, where it is thought natural and convenient for a girl to become her uncle's second wife—is somewhat changed by moving the relationship between the couple one degree further away: an Urubu prefers to marry the daughter of one of his cousins.

The kinship system can also be an abstraction if one forgets that relationship involves physical acts as well as mental ones. Like any process, its only justification is that it produces something new, besides maintaining itself: so that, while marriage may be the test of the kinship system, the proof that the test was necessary only occurs when a child is born. The bride and groom, and their families, are then related physically as well as socially; besides this, the couple then realise what mythology says they truly are: the man has become a father, the woman a mother.

The whole connection beween incest, marriage and children is a very interesting one, which the Indians' mythology deals with at some length though, of course, in a very symbolic form. The connection, in fact, is so hidden behind symbols that I only began to understand it when puzzling over a curious portent known to the Indians as the moon's vomit. During the process of turning a manioc tuber into flour, there is quite a lot of waste: the skin, the large fibres that run down the middle of the tuber, and other odds and ends, are all thrown away round the flour house. This refuse very occasionally is turned bright red by the action of a bacterium called *Micrococcus prodigiosus*, and it is this apparition which the Indians call the moon's vomit. As a portent of death, it is infallible: the first person who sees it dies.

When I first heard about the moon's vomit, I made a note of it as an odd and rather silly belief, and thought no more about it. Later on, however, Antonio-hu told me that at certain times during the rainy season the moon becomes quite red because, as he said, it knows that someone is going to die. Death, the moon, and redness must then, I realised, have a real mythological connection, which I could dimly understand—but why was vomiting linked with them?

First of all, the moon and redness. This was easy: the moon is the cause of menstruation, and redness indicates blood. I asked Antonio-hu to tell me the story of the moon, and he responded gladly. This is what he told me.

A long time ago, at the beginning of the world, the Indians were having a cauin feast in order to name a child.

These naming feasts begin at sunset, when everyone starts drinking; the actual ceremony only takes place the next morning, when the sun rises again. Night fell: there was a lot of drinking and singing, and under cover of the darkness a couple started flirting and at last went off together to a small hut, walled in with lengths of split palmwood, that stood at one end of the village. The man and the woman were actually brother and sister, though the man didn't know this; the woman knew, however, but she didn't care.

"I'm going to paint you," she said, "to make you look beautiful," and she took some genipapo and with it daubed the man's face, smearing it over his nose, his cheeks and his chin. She was a bad woman, for she might at least have said, "I won't paint you, you're my brother——" But she didn't. Then they made love together, they made love all night. She really was a bad woman. Why didn't she say, "No, you're my brother, don't suruk me"? But she didn't: she painted him, she slept with him, he suruked her.

In the morning the father of the child that was to be named called to the man—"It's morning, come and give my child its name. Come and dance my child for me!"

"Is it morning?" the man asked grumpily.

"Morning!"

So the man got up, washed his face, looked in his small mirror to see if his feather ornaments were on properly—horror! his face was all black, there was nothing white there but the whites of his eyes. Black black black! His sister looked at him, yes, he was black!

"Come on out!" his mother called to him from outside. But the man wouldn't, he hid his face in his hammock for shame. "I can't come out," he said, "my face is all black." Everyone crowded into his hut to look and, sure enough, his face was all black. "O woman," they cried to the sister, "why did you want to paint his face? Why did you make him black?"

The man was implored to come out and name the child, but he refused. So they found another godfather, who danced the child and named it in the proper manner,* and the feast continued till night fell again and the cauin had run out.

In those days there was no moon, and when the sun set the world was in complete darkness. When everything was black, the man suddenly came out of his hut with a great bundle of arrows under his arm, and rushed away down the path into the jungle.

"Don't go away," the villagers cried after him, "stay with us, don't run off!"

But he did, he ran off, and his sister followed him. He ran till he came to a small clearing in the middle of the jungle. He tightened his bow-string, and shot an arrow into the sky. The arrow went right up, hit the sky and stuck there quivering; then the man shot another arrow, which hit the butt of the first one. In this way he shot off all his arrows, till he had made a kind of rope of arrows hanging from the sky, and touching the ground. Up he climbed, till he reached the sky; and his sister came after him naked, having thrown away her skirt.

Three days later the Indians suddenly saw a new moon in the west. "Look at our brother," they cried, "he's turned into a moon!" And his sister was there too, following him; she had become the evening star. Five days after the moon had first appeared, all the women began to menstruate. None of them had ever menstruated before, because there hadn't been a moon, but now they all started. "Now I've seen a bad thing," they said to each other. It was the blood coming out.

This version of the story is not quite as clear as it might be. By stressing the bad behaviour of the man's

* See page 186.

sister, Antonio-hu missed the point of the painting incident: other Indians say that the woman suspected it was her brother who slept with her, but could only make sure by marking him with genipapo—a kind of Cupid and Psyche myth. She was urged to do this by her mother—"when a man goes into a carapua at night he should tell his name, so that the woman knows who it is," she said—and paint him she did, instead of just asking who he was. This was a kind of incest all on its own, for a man will die if he is painted by one of the women in his immediate family, unless she be his wife.

Mythologically, however, it was necessary that the man be painted by his sister, because otherwise the black marks on the moon's face could not be explained; and here another set of mythological facts must be considered. Indians paint themselves for two different reasons: they paint themselves red with urucu in order to vivify themselves, but when they want to be protected against the supernatural world they use genipapo. They do this especially whenever they drink cauin, in order to prevent the cauin from 'cutting their livers' as they say.

Now the brewing of cauin, as I suggested before, is the result of a kind of alchemical process, during which the matter of manioc is transformed into 'spirit'. It also has a direct connection with the moon, for in one odd but very fragmentary story which I heard, an old woman was warned against drinking a bowl of cauin because it was really the moon's faeces, which would do her no good at all if she swallowed them. The moon, therefore, can be understood as a presiding spirit over all magical transformations, for not only does it cause women to menstruate, but it is also, in some way, the genius of cauin. Besides that, it is itself a transformed man, a transformation effected by incest.

Why should incest turn a man into a moon? One of the most widespread of South American myths—a variant of it was told among the Tupinamba—tells of the flood that covered the earth at the beginning of the world, drowning everyone except for a brother and his sister who had climbed a palm tree. When the waters receded, they climbed down and, since there was no one else alive, slept with each other and so started a family.* Their children also had to marry each other; and from such incestuous unions a great many Indian tribes think themselves descended.

It is a great paradox that society should take as its origin an incestuous marriage which is the negation of everything society stands for. But here the Urubu story of the moon points the way out. The incest then committed did not create what we would call a social order, but a supernatural one: by appearing, the moon made women menstruate, and so brought into being a system of taboos and an awareness of the fact that acts have spiritual as well as physical consequences. Primitive man, it seems, orders his life magically before he orders it socially through such institutions as marriage and the kinship system.

This supernatural order, which the moon creates, is a dangerous one for man to enter. The moon is the patron saint of women rather than men, and the powers it symbolises are, mythologically speaking, female; it shines at night when the sun, symbol of the masculine ideal, has set. This opposition between the sun and the moon implies another paradox in the story Antonio-hu told me. It is Maír who is the real culture-hero, and when a child is named it is taken in the arms of its godfather who dances and presents it to the sun, not to the moon. Why then does the myth insist that it was at a name-giving ceremony that the incestuous act took place which gave rise to the moon?

* The connection between rain and incest is also apparent in the Indians' belief that whenever a tapir is killed it will rain soon after—tapirs being notorious for their incestuous unions (p. 82).

The question has already been partly answered by connecting the moon with cauin: it is impossible to hold such a ceremony without cauin, because men must first transform their bodily energies, by getting drunk, before they can transform their spiritual ones in honour of Maír. But besides this there is a certain confusion in the Indians' minds about the exact nature of the forces symbolised respectively by the sun and the moon. Take, for instance, Maír's particular animals, the jaguar and the snake. Maír's main symbol, of course, is the sun, so that one should expect the jaguar and the snake to exhibit sun-like characteristics. On occasion, it is true, they do so; but more frequently they are thought of as denizens of the underworld, the world through which the sun must pass at night on its journey from west to east. This underworld is mythologically a female place, as opposed to the sky which is the male place; jaguars and snakes, therefore, are as much emblems of the moon's supernatural order as of Maír's. Take, too, the tamandua anteater who, with his long nose, is the perfect embodiment of an Old Man of the Woods. Why are so many Indian tribes convinced that all tamanduas are female?

The Indians who live in Guiana tell a comic story that illustrates this odd reversal of mythological gender. One day, they say, the jaguar taunted the anteater for being a weakling, because he only ate ants. The anteater said it was a lie, he ate flesh all the time, it was the jaguar who ate ants; and to prove this he suggested a test, that they should both shit, and so demonstrate just what they had been eating. Jaguar agreed. "Let's start," he said. "Wait a moment," said Anteater, "we must do it with our eyes closed; then, when we've both finished, we'll turn round and take a look at the same time." Jaguar agreed. He closed his eyes and was soon finished—"shall we turn round now?" he asked. "No, wait," said Anteater, "I can't do this in a hurry." But he had already finished, and he was then exchanging his pile of faeces for Jaguar's. "All right," he said, "let's look now." To his amazement, Jaguar saw that the faeces under his tail were full of ant legs and carapaces, while it was plain that Tamandua, by the look of his faeces, had been eating flesh.

So the jaguar and the anteater change places; and, to show that this really does imply a change of mythological gender, the Gê Indians, who live to the south, tell almost the same story, the significant change being that the actors are two brothers called Sun and Moon. It is then not surprising that the Urubu story of the moon should be staged during a name-giving ceremony during which a child is presented to the sun; Sun and Moon are both transforming powers whose nature is never clearly defined, and who anyhow cannot work by themselves—each needs the other to be effective.

The story of Jaguar's humiliation brings us back to that curious identification of cauin with the Moon's faeces. The identification is made because both are the results of the supernatural processes symbolised by the moon, just as a child is the result of the physical process set into motion by marriage. Mythology is full of mentions of faeces presumably because they are the first things that a child produces by its own unaided efforts, and therefore are primitive symbols for everything that human beings create. One more myth, and this will be demonstrated; at the same time, the significance of the moon's vomit should become plain.

For some reason, perhaps because they have intermarried with other tribes, the Urubus speak of two culture-heroes. Maír is the one they nearly always talk about, but sometimes they speak of Anawira who, like Maír, came out of a tree, bearing manioc, the clay plate on which manioc flour is roasted, and fire. The tree he emerged from is called anawira, like himself, and it bears large, rather bitter nuts which can be grated and turned into flour, like manioc, and which the Urubus once made use of when they abandoned their villages and took to wandering through the jungle as nomads.

When Anawira emerged, there were already men and women in the world. Their food was rotten wood, and because none of them was provided with an anus at the time the only way they could defecate was by vomiting up what they had eaten. One day they went to see Anawira, and with true hospitality he made them a huge potful of shibé, which they drank till their stomachs were distended. One of the men drank so much that his bowels couldn't hold it all, so he began to vomit on the floor of Anawira's hut. "No," said Anawira, "don't do that," and he went to the edge of the clearing to cut a length of urucu-i liana. He sharpened it into a point, made the man bend over and thrust it between his legs, so cutting an anus.* The man was overjoyed: he rushed into the woods and didn't bother to squat down, he just lifted a leg and let everything come out, first the rotten wood, then the shibé.

"Now you know how to shit," Anawira told him; "you shit through your anus and not through your mouth."

"Yes, now I know, now it's good," the man said. So then Anawira took his stick of urucu-i and cut anuses for everyone, and for the women he also cut vaginas. The men had only tiny penises in those days, but they soon grew, because the women wanted them.

When Anawira had finished his work of cutting anuses, the Indians all went to clear a patch of jungle nearby, where under his guidance they planted manioc. They were like animals in those days, for they would make love right in front of everyone and Anawira had to tell them not to: love-making, he said, should be done in private.

The Indians were indeed like animals, especially before

* The myth here takes an everyday occurrence, and reverses it: for Indians clean themselves with a stick (rather than a leaf) after defecating.

they went to see Anawira. The rotten wood, which was their food, is said to be eaten by tortoises; and these are, like anteaters, Old Men of the Woods. Rotten wood, besides, is the stuff out of which Maír made the nomad Guajaja Indians—the Urubus, like all 'real' Indians, were made from growing trees. In those days, therefore, the Indians were hardly human. To become human, they had to be worked on by Anawira.

The piece of urucu-i liana which Anawira used as a knife is the wood from which fire-sticks are made. Two sticks are necessary to make fire; one is whittled into a point, and in the side of the other a small hollow is cut, with notches at each edge. This 'female' stick is laid on the ground, and the sharp end of the 'male' stick is put into its hollow and then twirled rapidly between the palms of the hands. After nearly a minute a thin blue smoke will rise from the point of friction, powdered wood-dust will fall down the notches on each side of the hollow, and at length a spark will fly out and ignite the heated dust. This way of making fire makes it plain why sexual intercourse should so often be symbolised by fire, and the penis by the fire-stick. It was thus quite in keeping with this symbolism that Anawira should use his fire-stick to cut vaginas between the women's legs; why he also had to cut anuses is perhaps not quite so obvious.

Fire, however, is more than a sexual symbol; it symbolises all creative activity. By the use of fire, meat is cooked, manioc flour is roasted, pots are baked, and men keep warm at nights when they sleep. It is the necessary condition, in fact, for all human culture which, in this story, is also symbolised by the great pot of shibé prepared by Anawira. But human culture, which uses fire in the process of transforming matter into food, is itself a process that changes man from an animal into a real human being. To achieve this, it must be allowed to run its course: once Man has eaten its food he must keep it inside him, digest it and at length expel it through his anus in an act of personal creation. To vomit it up is to reject it and to choose the animal and not the human condition.

When a man sees the moon's vomit, therefore, he knows he is to die because the transforming power itself has gone into reverse and thus denied that it is effective. Manioc, when it goes through the transforming process, ends up as flour, and its waste is thrown away; the moon's vomit implies that the person who sees it has been eating waste (just as men long ago ate rotten wood), and has thrown away the flour. This waste, when it is vomited up, is coloured red, to show where it comes from; for red is the colour of blood, and when it comes from the moon the man who sees it 'sees a bad thing': it is menstrual blood, which can turn a man yellow, make his hands shake like a dotard, and soften his bones till he can no longer stand up. The moon's vomit, when it appears, is the most fatal of all portents—little wonder, since its significance is so far-reaching.

CHAPTER FOURTEEN

Ghosts and Children

WHEN one is thinking about Indians, one always comes back to a few general concepts, such as the distinction between the jungle, the home of animals and ghosts, and the safe, ordered world of the village. Ghosts, as far as I know, never manage to enter a village on their own; the supernatural there is controlled by society, and only appears in the form of portents such as the moon's vomit, or the owls that hoot whenever a man is about to die—the cupi anyang owl, for instance, that cries 'wa! u! wa! u!' which in Urubu means 'man! eat!' To see a real ghost one must go into the jungle.

Indians often see ghosts, and it is not surprising that they should, for the jungle is full of suggestive noises. Hummingbirds whirr with great speed and precision between the branches, there are sounds of twigs falling, ripe fruit dropping, parrots squabbling in the tree-tops, even the sound of ants moving over dead leaves on the jungle floor. If one is by oneself these noises may gain a menacing intensity, so that one imagines that Things are waiting in the shadows; and, at night, even the company of a friend may not be enough to quieten one's apprehensions.

Pari and Mbeiju, for instance, went off one day on a lengthy hunting trip with their dogs in the hope of killing a lot of game. They were unlucky, as usual: they came back three days later with only two paca and four or five agouti, complaining that they had been haunted. A perfectly ordinary jaguar had frightened them first, by roaring angrily at them from the bushes, and then they had heard Anyang, the spirit of the dead. Anyang had come near their camp moaning to itself, meh meh meh, and had gone away only to return at midnight and shout at them; Pari had been so disturbed that he got out of his hammock and made a huge bonfire, near which he squatted till dawn, bow and arrows to hand. Next day they moved their camp to another place, but Anyang had followed and again shouted at them in the middle of the night. Pari didn't go hunting again for quite ten days.

It is quite common for Indians to hear ghosts. Paijé was once gathering bacuri fruit close to Antonio-hu's village, when the rain began to fall. The wind got up, and suddenly she heard the noise of someone keening, as people do at a death; then the storm broke, the wind roared and the rain flooded down, which put her into such a panic that she ran all the way home. This seems to have been a wind-ghost: but there are others whose physical cause is less easily understood. On another occasion Paijé and her family were coming back at night from a hunting trip when she heard Anyang groaning by the side of the path. No one else heard it, so she just stopped where she was and gently hummed a shaman song, in order to drive it away. Anyang groaned; she sang, and at length exorcised it.

To see a ghost is more dangerous; and it is essential then not to talk about it for a considerable time afterwards, lest you fall ill and die. Alishandre once told me of a very odd thing he'd seen in the rapids above Canindé. He was shooting fish at the time, and in one of the shallower pools he saw a man, quite yellow, lying face-down on the bottom, covered with small fish. He was quite sure this was not a drowned man, for it had been seen many years ago in exactly the same place by a woman; but she had seen this spirit man, who it seems was of a surprising beauty, lying on his back and looking up at her. Some days later she caught a slight fever, and thinking her vision might have something to do with it, told her husband what she had seen: a fatal mistake, for she then died. Alishandre only told me of his own experience because it happened such a long time ago.

This keeping quiet about a ghost you may have seen is important in order not to turn the potential relationship between you and the ghost into an actual one; for ghosts usually only appear to those who are ill as symptoms of that illness, and portents of their end. It is, however, quite possible to the Indians' mind that one should see a ghost for no particular reason—it may be on quite other business when it runs across you—but if you speak about this openly, the ghost may take it into its head that you are the object of its search, and will forget its original purpose and attack you.

If you are ill already, however, and you see a ghost, there is little hope for you. Pari's first wife, for example, knew she was to die because a ghost appeared to her. She had been very ill, and Pari had taken her to Canindé to ask for a medicine; there she recovered slightly, but on her way back home she saw two ghosts and became extremely frightened. One of these ghosts was an agouti in a bush near the path, the other a paca: and she knew they were ghosts because they suddenly disappeared, no-one else had seen them, and the dogs when urged to pick up the scent were quite bewildered and cast vainly about: there was no scent. The woman decided that the paca and the agouti must have been the ghost of her brother, already dead, come to tell her it was time to join him in the sky. A week afterwards she saw the moon's vomit on some refuse near her hut, abandoned all hope and died.

Sometimes these portents appear to Indians who are not ill, but just old. One man I knew, who was always known as Tamuĩ, Grandfather, was picking lemons in a deserted garden when he heard a dreadful ghostly shout behind him that sent him running back home. For his age—he was well over eighty, perhaps ninety—he was extremely active, bringing more game into the village than any of the other men; and though he had fallen ill the year before, he was to all appearances perfectly sound. (I had actually treated him in his illness; he had a great abscess on his thigh, caused by a snake-bite twenty years old—the poison had affected the bone, and the abscess was the result. But the abscess, with the aid of sulfa drugs, had cleared up, and he had recovered.) Tamuĩ, after hearing the ghost, retired to his hammock and died peacefully some days later, not because he was ill, but just because it was time for him to die. Who the ghost was that announced his death, none of the Indians could tell me: it was just a ghost, shouting for no particular reason. Tamuĩ, however, knew otherwise.

I myself suspected that the ghost was Timakanã, the leg-bone ghost (or perhaps one should say spirit) who had been seen, a long time ago, by another Indian near that particular lemon tree. This Timakanã had several haunts, one of them in a swampy piece of ground not far from Pari's village, another near Antonio-hu's village: and this last haunt was the scene of a sinister story told me by Antonio-hu.

"One day," said Antonio-hu, "a man went into the jungle to get tortoises. Off he went! He picked up several, made camp, cooked one of the tortoises and ate it, and was just going to sleep when he heard a noise there in the distance, coming nearer and nearer—ti-ye-wo! went the noise, ti-ye-wo! Like a tin," Antonio-hu said, picking one up and beating it with a spoon.

"Ti-ye-wo," I said, "what's that?"

"I don't know," he replied, "I don't know what it is, no one knows. One doesn't know what thing it is that makes the noise. It's tall, and its knees are swollen—there seems to be a calabash round each knee. But one doesn't know! When it walks, it limps, and its knees beat together: they go ti-ye-wo, ti-ye-wo! Then there's another lot of calabashes round its neck, which shine with light . . . What is it? I don't know. It's tall . . . Timakanã!" he suddenly cried, "that's its name, it's Timakanã that walks ti-ye-wo . . ."

Antonio-hu continued his story. " 'HU!' shouted yandé ramuĩ." (Whenever Antonio-hu tells a story, he calls its hero yandé ramuĩ, our forefather—usually because he cannot remember the man's real name. This particular yandé ramuĩ, however, was the brother-in-law of a woman living in Antonio's village.)

" 'Hu! Hu! Hu!' the man shouted in fear. Timakanã was right by him now, in the bushes by his side. 'Hu-hu!' It didn't budge. Yandé ramuĩ got really angry. 'Go away or I'll shoot you!' he shouted, and he took his bow and shot an iron-headed arrow into the bushes. He shot one, he shot two, but none of them hit Timakanã, they all missed. So he took a blazing log from the fire and rushed at Timakanã, rushed at it with a great shout in order to drive it away; he drove it off ten, fifteen yards. But when he came back, there it was, sitting by the fire. Mai-té!" Antonio-hu exclaimed. (Mai-té: literally, Thing-indeed: that is, a wonder.)

" 'Ti-ye-wo!' went Timakanã's knees, as it bent them. Yandé ramuĩ was now really scared. He got a frond from an assaí palm, dried it carefully over the fire, then dipped it into the flames and, when it blazed, rushed again at Timakanã, driving him away for thirty, forty yards. Maí! But when he came back, there was Timakanã, sitting by the fire. Eh, yandé ramuĩ was frightened! He spent the rest of the night on the other side of the fire, standing in the shadows, not daring to sit down or to close his eyes. Towards morning, Timakanã got up and went away a few yards.

" 'What's it doing now? is it going to kill me?' yandé ramuĩ asked himself. Timakanã came back, and was edging round to where the man stood, when the saí birds began to sing, announcing the dawn; so it went away and didn't come back.

"Yandé ramuĩ was tired, tired. His eyes couldn't see anything, he was so frightened; he could hardly stand up. He undid his hammock, put it over his shoulder, and went wandering off back home, trailing his bow behind him.

" 'Where have you been?' his wife asked, when he arrived late in the afternoon.

" 'I've seen Timakanã, he stood by me all night and he's eaten my soul,'' said the man. He fell into a hammock and stayed there five days without eating anything, without drinking anything: his soul had been eaten up. Then he died.''

If you should meet Timakanã in the jungle, Antonio-hu went on, you must remember to be polite. There was once a hunting party of five or six people, who had killed a lot of peccaries and had to stay the night in the jungle while they smoked the meat; and Timakanã approached them, but didn't eat their souls because they treated him well. 'What are you doing here?' they shouted at him, 'why look at us? We came out together, we've been killing peccary and now we're eating peccary. Go away, go away and kill peccary for yourself!' Timakanã went. It's only when you shout at him rudely, insult him, shoot arrows at him, that he gets angry and eats your soul.

He eats your soul: this describes accurately just what a ghost does to an Indian who is so spiritually disturbed that he cannot protect himself against it. But an even worse fate can befall one: that of having one's soul eaten and yet remaining alive as a kind of zombie. This is the speciality of Anyang.

Anyang is one of the nastiest of the jungle spirits. It is the spirit of the dead, so that it sometimes appears as a single figure, sometimes as many, both male and female.

The Anyangs go about perfectly naked, the males with their prepuces untied, and they have a revolting dance where they cavort in a circle, the females flapping their dugs up and down with their hands, the males doing the same to their enormous penises. Their hair comes down to their waists, and is all tangled; their skin is green, the colour of decaying corpses, and there are no bones in their bodies. Antonio-hu told me of a time many years ago when there was a horde of Anyangs on the Capim river, where the Urubus then lived; they attacked a family there, kidnapped a small child and were just carrying it off when the father of the child seized a horn and blew it. This frightened the Anyangs so much that they dropped the child and ran howling into the jungle, tripping over roots in their flight.

It is dangerous to antagonise Anyang, even if you are only doing it in self-defence. A few weeks later the father was hunting in the jungle when he was attacked by the horde of Anyangs, who ate his soul and turned him into one of themselves. For a year he wandered through the jungle, living on berries; quite dumb, his hair growing long and matted, and his skin turning green with dirt. His whereabouts were at last discovered magically by a shaman who then cured him with tobacco smoke, and told him that he would be all right as long as he never painted himself again. But his wife wanted him to look handsome at a cauin, and put urucu on his face: he died almost immediately.

Urucu, being the colour of blood, is an invigorator, that no man may put on his face if he is undergoing seclusion. At such a time he is pretending to be dead, and to use urucu would be as bad as drinking poison. The man, of course, was not in seclusion, but in a much more dangerous state: he was a living corpse.

Running throughout these stories is a thread of sense, for ghosts rarely appear without good reason: they are parts of the soul, and symptoms of an internal dissension. The soul itself is called *ang*, which literally means a shadow. Like most mythological terms, this is full of meaning—it implies that a man can only see his soul when the sun (or rather, since we are speaking mythologically, the Sun) is shining. Indians are always putting themselves into relation with the Sun, that is with Maír and consciousness in general, but they do not make the mistake of thinking that the soul is actually the sun ; the soul, like the shadow, is a projected image, and a negative one at that. When it meets a ghost which is too strong for it, the soul loses this sense of relationship and is 'eaten'.

Anyang and Timakanã are a kind of super-ghost, ghosts that are common to all Indians, for they usually appear as the consequences of definite states or actions. Anyang often appears when a man has a new-born child at home; Timakanã when a man goes gathering tortoises, or if he has killed too much game. These things are related in a curious way. Anyang can be left to one side for the moment; Timakanã is the more important figure, though he is not a member of the usual canon of spirits recognised by Tupi-speaking tribes. He belongs, instead, to a different mythology, presumably the same one from which the Urubus got their second culture-hero Anawira. But Timakanã has a close connection with that other jungle spirit, Curupir. Curupir (literally, Rough Skin) is, the Indians say, the 'owner' of the jungle—its presiding spirit. His feet are turned round at the ankles, so that his footprints point in the direction he's just come from—a way of deceiving those who try to escape him by running away in what they think is the opposite direction. Now this, which is a classic trait of witches the world over, seems to be linked in Brazil with the giant anteater, whose front claws are so long and curved that it has to walk on its knuckles, the claws pointed backwards. The other main characteristic of the anteater is its long nose, because of which it often figures in myth as an Old Man of the Woods; and this phallic character also belongs to Curupir, who has two enormous testicles in a pendulous scrotum. With these he announces his presence in the jungle, swinging them with great force against tree-trunks and especially against the large, buttress-like aerial roots so many jungle trees possess, which resound magnificently when beaten.

Besides this, Curupir is known by two other traits: his eyes, which at night are said to blaze like small suns, and his close relationship with tortoises. Both these traits show his likeness to Timakanã, who emits light from the calabashes that cover him, while the calabashes themselves seem to be a euphemism for tortoise carapaces. The most obvious connection between them, however, is their 'ownership' of the jungle. If any hunter kills more game than he can possibly eat or take home, Curupir at once confronts him and either makes him *panem*, unlucky, or else kills him by supernatural means. Timakanã, it seems, when he approached the hunting party that had killed a number of peccary, was wondering if they hadn't killed too many. If they had, then not all their politeness would have helped them: he would have eaten their souls.

Curupir is thus the chief of the Jungle, and as such he stands in opposition to Maír, the chief of the Village. The difference between these two figures is strange, for it is almost as if one was the mirror image of the other. If Maír walks forwards, then Curupir with his twisted feet walks backwards: Maír shows himself as a sun, during the day, and has no penis, while Curupir usually appears at night when his blazing eyes are seen to the best advantage, and sports enormous genitals. Their opposition to one another is well seen in a curious reversal of custom whenever Indians eat deer in the jungle. The whole business of cooking and eating deer is surrounded by taboos, which have been mentioned earlier, because deer are magical animals or incarnated ghosts. To bring the carcase of a deer into a village is to confront Maír with everything that he fights against: taboos are therefore necessary in order to control and isolate this war of opposing spiritual forces. In the jungle, however, Maír is absent, and the hunter is on the same footing as the thing he hunts: taboos thus no longer apply.

But it is the tortoise which provides the most interesting connection between Maír and Curupir, being the real intermediary between jungle and village. Tortoises are a kind of animated fruit, for they can be picked up, as one gathers bacuri, either from the beds they make for themselves among the leaves, or under the trees where they come to eat fallen petals and fruit. Except for manioc flour, tortoises provide the food an Indian most often eats, and one of the only foods he can eat in seclusion.

There are two kinds of tortoise: carumbé, with red scales on its legs and head, and a dark brown shell; and yashi, smaller, with its shell and scales a creamy-brown colour. Carumbé, whose name seems to mean *food-thing*, is a heroic character in mythology: he lives in a large hole in the ground, singing and shaking a maraca made from a jaguar's skull. From time to time he comes out and by various ways, often very crude, succeeds in killing another jaguar or two: then he retires again, with a new jaguar skull to use as a rattle. The tortoise's virility also shows itself another way, for its penis is used as an aphrodisiac. I was made to eat one once, after I had carelessly said that I was not particularly longing for a woman at that moment. Pari, who had asked me, and who was longing for one, thought there must be something wrong with me; so he got Chico to open the tortoise we were going to have for supper, and grill its penis over the fire. It was a nasty sight when it was cooked, greenish-black in colour; I took a couple of dubious bites, chewed hopelessly for five minutes, and then gave up. Pari asked from time to time whether I felt any different, but I detected no urgent signs, and this he put down to my having eaten a banana just before, well known for its mollifying effects.

Yashi seems to mean *white calabash*, and unlike carumbé it does not figure in any myth. But it is yashi that is eaten by Indians undergoing seclusion, and not carumbé, because its scales are not red but white: for red is the colour of life, and a man secludes himself in imitation of death. Because of this it may well be that the word *yashi* is cognate with *yahi*, the moon: the yashi tortoise thus being the living symbol of that lunar force which, by bringing about menstruation, made seclusion and the eating of tortoise flesh a necessity.

The spirit of the dead, of course, is Anyang; and it is interesting that he should be so fond of kidnapping children, for this fondness is also to be seen in Curupir. Curupir (and Timakanã) appear most often to people who go out into the jungle looking for yashi tortoises in order to have something to eat when they are in seclusion; and the most common type of seclusion is that of the couvade,* a period of some six months that follows the birth of a child.

When a woman is about to give birth, her husband takes to his hammock and imitates her lying-in. They are both in a state of seclusion, and so they take to the usual diet of shibés and yashi flesh. A week after the birth, the couple can get out of their hammocks and walk about in the village, but the man cannot go hunting, nor can the woman work in the manioc plantation. The diet becomes less restricted as time passes: the flesh of smaller game is allowed, but not of pig, deer, tapir or birds; and all sexual intercourse is still forbidden. Should these taboos be

* The couvade is properly a man's ritual imitation of his wife's confinement, but for convenience I am using the term to signify the period of seclusion that follows child-birth and the restrictions that both husband and wife must then observe.

broken, the child will very likely die; and the couvade only comes to an end when the child is some six months old, and receives a name.

The Indians take the couvade extremely seriously. It is true that some fathers, sickened by a diet of tortoise flesh, are occasionally tempted to eat a juicy piece of tapir, but they rue their lapse bitterly afterwards—as did Yawaruhu, whose young son died shortly after he had accepted such a forbidden titbit. His sorrow expressed itself in a kind of angry melancholy which his wife made worse by refusing to sleep with him for having been such a fool. But it is not just in order to protect their children that the Indians observe the couvade: they do it also to protect their souls. Tero, for example, once fell unaccountably ill and turned quite yellow. His yellowness was a sure sign that his illness was not merely physical, and that he had broken some taboo: and so it turned out. A married woman, with whom he had been having an affair, had recently given birth to a child. She and her husband had gone into seclusion and this protected them and the child; but it did not protect Tero who, by Indian logic, had contributed something to the making of the child, and was therefore under its influence. He was immediately made to observe all the rules of the couvade; in addition he was washed in a magical decoction of manioc leaves and water, and his legs were scratched with a sharp tooth till the blood flowed. This is not a usual feature of the couvade, but was necessary because Tero had broken a taboo and was in spiritual danger.

The relation between father and child, therefore, is a two-way one, the child laying as much claim on the father as the father does on it. The reason behind this is simple enough: a child, to an Indian, is the material proof of his own soul. Until he is a father, a man's soul is just a shadow or ghost, and like a ghost it is always wandering about, in night-dreams or day-dreams, looking for some material object in which to embody itself; only then can it be seen, and thus controlled. The soul finds various resting places, such as the chief of a village who is one of the obvious mirrors by which Indians see themselves—but it is only when it comes to rest on a child that it is best seen, for a child (like the soul) is the fully human creation all men and women are capable of.

There is, of course, one weak link in this argument: the fact that it is women who bear children, and not men. The couvade is a cunning way of getting round this, for the father is then trying to make the child his own in the most obvious way he can think of, namely by imitating the woman who gives birth to it. For six months this effort at spiritual creation continues, till the child is given its name and 'wakes up'; before this happens it is said to be 'asleep', that is, spiritually unformed.

The child is given its name at a feast to which all the surrounding villages are invited. The feast begins at sunset, when everyone is given a small sip of cauin; then the big bowls, which hold a litre or so each, are brought out and the drinking begins in earnest. The low, grumbling songs in praise of the cauin are sung, telling how cauin is made from manioc pancakes, and what its effects are; and the whooping bird songs are also popular, especially that of the yapu, Maír's favourite bird (for its yellow tail-feathers form the main part of the ceremonial headdress men wear in honour of Maír). At dawn everyone is invited to go to the largest hut in the village and witness the ceremony. The men sit on their hammocks, inside, the women on specially woven palm-leaf mats outside. The naming of the child is done by a *pai angau* or godfather, and usually two men prepare the feast together and name each other's children: they are often brothers-in-law or cousins. The godfather, dressed in all his feather ornaments, takes the child in his arms, and if it is a boy, puts a headdress on its head (girls may not wear headdresses) then dances with it towards a small fire which has been lit in the east, and towards the sun which is then rising. As he dances, he blows the hawk-bone flute hanging from his feather collar, and continues to dance and blow till the child stops crying. Then he gives the child back to its parents, stands up and announces its name, repeating it over and over again while he points in turn to every person who has witnessed the ceremony. Everyone repeats the name as they are pointed to, and there is then no excuse for forgetting it.

Sun and fire, of course, are both emblems of Maír; the godfather in his feather regalia is himself a Maír, and his dance shows Maír in his rôle of warrior, fierce as a hawk. The child is thus being presented to the world as a potential Maír, a claim which has to be made publicly in order that it can come true—just as, in the contrary sense, a man keeps private his dealings with ghosts lest they attack him. The father, of course, has already made a claim of a different kind during the couvade. The name-giving feast that brings the couvade to an end shows that this claim has already come partly true: for since the couvade entails seclusion, and seclusion represents the state of being dead, the man's emergence from seclusion implies his rebirth. The feast in fact celebrates the man as a father as much as it celebrates the child as a Maír: which is as nice an identification of child and soul as one is likely to find.

So it is no wonder that when ghosts attack people they either eat their souls or kidnap their children: the effect in the end is exactly the same. Nor is it strange that the tortoise, Curupir's familiar animal, should be eaten by men in seclusion who are preparing to turn themselves into Maír's. A tortoise, with its hard shell that arrows cannot pierce, is a kind of Maír itself; and it also symbolises the fact that Maír and Curupir, though standing for very different things, are at bottom but the two sides of the same coin—as the following story shows.

A young man went off into the jungle one day to gather some yashi tortoises. His wife had just given birth to a child, and yashi was the only meat they could eat. He searched all day for tortoises but didn't find any: every tortoise bed that he saw was empty, and its occupant was nowhere around. In the evening he heard a man bellowing some distance away—"I've got an anhinga snake round my ankle, come quick and take it off! quick quick!" A lie! it was Curupir, and he hadn't any snake round his ankle. The young man went a little closer, and again Curupir shouted. There was Curupir, standing under a tree, surrounded by tortoises which were bound on to poles, ready to be carried away.

"What are you doing?"

"I'm looking for tortoises but I haven't found any, nothing but empty beds."

"That's right, I've been in front getting them for you. Come to my house and spend the night there."

They went to Curupir's house carrying the tortoises—Curupir had given the young man a lot, but they were all ugly ones, small ones, with twisted legs and knotted shells. They came to the house, but it wasn't a real house, it was a hole in a tree; to the young man then, however, it looked like a house. Inside was Curupir's wife, who broke open the tortoise for the young man's dinner. She broke open a small ugly one, though the house was packed full of big ones.

"Sleep here tonight," said Curupir. They slept there that night, and in the morning Curupir said, "I'm going hunting, going to kill a mahau deer for you to eat. Stay here with my wife." A lie! Curupir was hoping to eat the boy for dinner that night. He took his wife aside and told her to take the young man out gathering naja nuts, and kill him. Then he went off.

"Come and get naja," said the woman.

"Let's go." They walked to a stand of naja trees, every tree a giant.

"Come here," said the woman, "I want your penis!"

"No, no, I've a young child at home, I can't suruk you."

A monkey was there, laughing in a nearby tree. "Eh!" said the monkey, "come here! What are you doing with Curupir's wife?"

"We're getting naja."

"No, she's going to kill you. Don't let her climb the naja palm, you climb it."

So, despite the protestations of Curupir's wife, he climbed the tree. There at the top were great clusters of nuts. "Mind you stand well underneath and catch the clusters as they fall," the young man called down, "so as not to scatter the nuts over the ground." The woman stood well underneath, spread her arms, and the young man cut the stem of the cluster . . . wuhuhuhu . . . umph! Down it came, just missed her.

"Catch it!" the young man shouted, and cut down another cluster. Down it came, right on top of the woman, right in her eyes, and broke her neck. All the monkeys in the trees started laughing, because she was dead.

"Quick," said the monkey, "take the body back to the house, make a fire, boil it."

So he cut up the body, cut off the head, cut out the carapua, made a fire there in Curupir's house, and put the flesh in a pot to boil, with the carapua right at the bottom.

"Quick!" said the monkey, and the young man ran away, back to his village.

Curupir returned, went into his house, didn't see his wife. "Eh! where are you?" No answer. "She's gone for a bathe, she's a good wife, she's killed that young man and put him on to boil. I'll eat him." He started eating the flesh, he took some flour and ate that too, then he thought, 'Where's the man's penis, where are his testicles?

I want them.' He searched about in the pot till he drew out the woman's carapua.

"Aie! the man's killed my wife and I've eaten her. Aie!"

So Curupir got his come-uppance: but it was a near thing. "Never," I was told, "never go hunting when you have a young child at home, or you'll see Curupir."

The point of this story is made clearer when one knows that there is an enormous palm tree in Maír's garden, the cousin to the one under which Curupir's wife was killed. Maír has a daughter, who can only be won by a man climbing to the top of the palm—it takes him three days, it's so high—and cutting down two clusters of fruit. Maír's daughter picks them up from the ground and makes them into a drink, which she gives to the man when he has climbed down the palm, just as girls bring their husbands a shibé as part of the marriage ceremony. Maír's palm was a bacaba, whose fruit makes a pleasant drink; Curupir's was a naja, whose fruit are eaten by rodents and especially by the opossum, the shaman's familiar animal.

The young man, just become a father, should have remained in his village in his secluded state of becoming a Maír. Instead he went into the jungle and thus tempted all the forces there to undo the soul he had been making for himself in the form of his child; tempting Curupir especially, since he was hunting tortoises. What should have been a propitious event then immediately reversed its values and became sinister: the naja took the place of the bacaba palm, and Curupir's wife that of Maír's daughter. It is easy to become a father, but difficult to be one.

CHAPTER FIFTEEN

Shamans

It is a pity that the Urubus no longer have true shamans among them. The shaman is a medicine man, who can cure diseases and see into the future, and can restore a man's soul should it have been eaten by a ghost. The shaman can do these things because every disease of mind or body is thought to have a spiritual cause, and can therefore be cured when the spirit responsible for it has been called into the village and magically constrained. The spirit is constrained by allowing it to possess the shaman, who must however remain in control; he, as it were, provides a body for the spirit to inhabit and to act through, but is strong enough not to let it turn him into a zombie. The shaman brings on his possession partly by singing the songs characteristic of each spirit, and dancing their particular dances, and partly by inhaling unbelievable amounts of tobacco smoke from huge cigars. The smoke not only helps to put him into a trance, but is also used as a fumigant when a disease is being expelled.

But the Urubus have no true shamans, probably because of a historical accident: the two brothers who came from the Capim and founded the new tribe to the south of the Gurupí were both warriors and not shamans, so that the magical knowledge, learnt only after a long apprenticeship, was lost. They remembered, of course, some of the more obvious shamanistic techniques, such as the use of tobacco smoke as a fumigant, but those who practised such techniques were not known as payé-té, true shamans, but as payé-ran, pretence shamans. Nowadays they have a number of these pretence shamans, who try to increase their knowledge by visiting the Tembé Indians at Jararaca, where true shamans still exist. Toí was one of them: he paid for his apprenticeship with a great bundle of arrows and, he confessed to me, he nearly died because he wasn't allowed to eat anything. "For that many months," he said, holding his hands up before his face and spreading the fingers again and again. "I ate nothing but tortoise meat."

Chico laughed. His father, who lived at Jararaca, was a true shaman, and had begun to teach Chico the elements of the shaman's art: but Chico hadn't bothered about going on a diet, he ate everything he liked. "I ate everything," he told me, "but then I didn't learn anything! A whole year I spent learning payé-té songs, and I still don't know them properly."

Toí, alas, didn't know them either. The text of the songs, being in a difficult, archaic Tembé, always eluded him, as it did all Urubus; he had learnt nothing but the tunes, which he would bawl out with as much assurance as he was capable of. It seems, however, that the old Urubu shaman didn't bother with songs, nor did he smoke his tobacco, as the Tembés do, in a cigar made with a wrapping of tawari bark, he smoked it in a clay pipe. In addition, he had but one familiar spirit who possessed him, that of the white scissor-tailed hawk or tapé-tapen. The shaman, smoking his pipe, would fall into a trance, tapé-tapen would descend, ask what the matter was, and give the relevant advice on which the shaman, when he had come to, would act. But it is difficult to get exact information on native Urubu shamanism, because it has not been practised for so long. Some say that the last shaman lived four or five generations ago; a notion Antonio-hu strongly objected to. There had only been two shamans, he said, who were born at the beginning of the world and went up into the sky to escape a universal fire. These mythological shamans never left their knowledge to their children because, he informed me tartly when I asked, they never married; nor would they have been much use to their wives if they had, for their penises were no larger than a child's. (To my great surprise Antonio-hu, when he told me this, suddenly became wistful and wished that he was like a payé-té, and that his penis wasn't so large. Usually his wishes were in the quite opposite direction.) These two shamans have much in common with Maír, and it is likely that they are no other than the miraculous twins born to his wife (Chapter 17); anyhow, the Urubus are thankful that they had no issue, and that there are no payé-té alive today. Stories of magic may be fun, but magic itself is not pleasant when it is misused; and misused I am almost sure it was, for among the Tembés the tapé-tapen bird, like the vulture, is the spirit of sorcery and not of healing.

Chico, though he was no shaman, yet knew enough to be contemptuous of Urubu pretensions in the matter. Pari told him how, when he was at Jararaca, a shaman had given him two *caruwas* to hold—spirit tokens, that can either cure or cause disease—and that he had seen a man sing shaman songs immediately after eating tortoise meat ('that's bad,' said Chico) and produce a butterfly caruwa.

"Butterfly!" said Chico indignantly, "butterflies don't have caruwas. Butterflies just come along, drink macasser (sweet manioc soup: the shaman's traditional food after he's been singing); they come when you sing just because they come, they don't *bring* anything."

But Pari, indifferent, was making the butterfly caruwa; a long piece of thread, with a pointed sliver of wood at either end, which he held by the middle and dangled before us.

"That's how it was," he said, and hung it up at the head of his hammock.

"They don't know," said Chico, jealous and protesting,

"they don't know! It's bad for them to sing when they don't know, they die; the patient dies, the singer, people die that way."

Terepik told how she had a headache once, and Antonio-hu came and sang over her, and took out the caruwa—just one song, and there! the headache went.

"It's no good," said Chico again, "if you sing that way it's dangerous, the illness goes for a bit but then it comes back worse than before."

Terepik didn't think so: her headache hadn't returned. Toí announced that he was going to sing that night, and Chico laughed scornfully. "He doesn't know how!" he said to me. "Tonight, I won't sing at all, I'll just listen."

By the evening, when everyone in the village had come crowding into our hut with their hammocks, which they slung here and there between the house-posts, Chico thought he'd sing three songs for them perhaps, just because they didn't know how.

There was a small pile of tawari cigars on the table, which I'd provided from my tobacco; made from hard tobacco sliced fine and rubbed between the hands, put into a thin sheet of tawari bark which is then rolled up tight and tied with narrow strips of itself. One such tawari, a foot long and half an inch thick, was lit and given to Chico, who started humming to warm himself up; then he turned and said to me, "I'll sing six or seven songs tonight for them all." He sat in Pari's hammock, with Pari at his side, the rest of us, men, women and children, each with a tawari, sitting or standing where we could.

After humming the tune to himself a few times—all Tembé tunes were pleasing and elegant, some sounding remarkably like Gregorian chants, in contrast to the whooping or groaning songs of the Urubus—and beating time with the maraca Toí had lent him, Chico started singing properly, and soon was on his feet, singing his loudest. The others, not knowing the words, all hoo'd and hay'd as best they could, supporting the tune while Chico started on his cigar. He played with it at first, crumpling the end slightly between his fingers and blowing, so that a stream of sparks went up into the roof; Toí at once followed suit, with Tero playing the fool with his cigar, using it like a hose. Then Chico started to inhale—no dainty puffs, but taking the smoke with great sucking gasps right into his lungs, working his shoulders like bellows to get as much smoke in as possible in the quickest time, and then out again, suck blow, suck blow, till I was sure he'd have to take a mouthful of pure air in order not to suffocate; on he went without even a cough. Admittedly, it's not all smoke that one breathes in: the mouth isn't clamped to the end of the cigar, but a small space left open which lets the air mix with the smoke, thus letting the lungs take in a great volume at great speed. But it must take a lot of practice: I cough after a couple of such cruel lungfuls. A good shaman, however, will smoke five or six tawaris eighteen inches long in one night, and while he'll do quite a bit of coughing he'll sing till dawn without losing his voice.

Chico sang quite well, for a young man: he didn't yet have the resonance his father had, which makes listening to Tembé songs a pleasure even when they go on all night. But the others were a bar behind him all the time, and when they couldn't find the tune they'd sing *hé hé* as a stopgap in a neutral tone of voice; while the women, like dreadful oboes played windily in their top reaches, sang —whatever the song and whatever the place in the song —a guttural yodel of four notes, always just out of key. Chico was soon drowned by all this noise, by the women's descant and the men's roaring whoops and *hé hé's*.

After every other verse or so, Chico would sigh exhaustedly—ehh! ohhh!—and take another twenty lungfuls of tobacco smoke. Then—if he wasn't caught by a fit of coughing, which came on him later—he'd start immediately into the succeeding verses. Ehh! Ohhh! the other men round him sighed, just to be in things; filled their lungs with smoke, roared out the tune, and blew showers of sparks into the air. It was really a remarkable sight, since it was a very dark, moonless night, and my paraffin lamp had been put out: all one could see was a large number of burning cigar ends, glowing and fading, sometimes emitting showers of sparks, and from time to time illuminating some dim, ruddy face anxiously sucking. The smoke filled the air, curling slowly out of the hut, and the noise was magnificent.

Chico, drunk with tobacco smoke and also, doubtless, with too much oxygen, started groaning deep in his chest till he left the ordinary songs that he'd been singing up till then, and started on a true shaman song, to summon Mikur the opossum from the jungle to aid him in the cure he was going to do. Soon he fell down on his back on to the ground, where he lay groaning miserably: the others all stopped singing, which they shouldn't have done, puffed idly at their cigars and chatted a bit; at length they started singing again, urged by the women. Four minutes later Chico was on his feet once more singing louder than ever, possessed by the opossum spirit. One of the children had a pain in her side, and while everyone continued singing, Chico palped the place, blew smoke over it directly from the end of his cigar, blew some more from his mouth. Then he sucked the place several times, after having filled his mouth with smoke, and vigorously blew into his closed fist, which he then flung away from himself, opening his fingers at the same time, to get rid of what he had sucked out. Finally he filled his mouth again with smoke, and blew it through his hand on to the place, and was finished. He showed what had been causing the trouble—a rather nasty little object, I think a piece of meat he'd kept in his mouth since supper.

Then he started giving a strengthening caruwa to Tero and Toí and Pari. He made them blow their cigars into his cupped hands, which he held in front of his face: then, with the aid of his hands and a fearful coughing vomiting noise, a kind of chesty barking belch, he blew the caruwa into their bodies. When he'd finished that he sat down to relax for a few moments, and so did everyone else, but he soon got up again, and started pulling at the men's arms and heads, pulling the skin, pushing it, kneading the flesh—to adjust the caruwa properly, he told me afterwards. At last the party broke up, but not before Chico had warned Tero to be careful what he ate in the future. He couldn't eat anything that had a strong smell, or the caruwa would leave him and, in leaving, hurt him: he couldn't even eat tortoise, because the caruwa might smell the shell burning as it was being grilled. And he couldn't sleep with a woman, that would be fatal.

Tero took his caruwa quite seriously for a couple of days, and refused to eat some peccary that was brought in because it had a smell. Chico tried to give me a caruwa: he put a piece of charcoal into my wrist, holding it first between the palms of his hands, then blowing on it, rubbing it on my wrist, blowing some more (when he took it into his mouth) and then slid one hand over my wrist and rubbed in the charcoal that wasn't there. I refused to let him take it out, which made him anxious, and for a couple of days I was doing all the wrong things. When I then complained of a stomach-ache from eating too many bacuri, Chico at once begged me to let him take out the charcoal since, he said, it was that which was causing the harm. When I laughed, he grabbed hold of my hand, made a pass over my wrist with his other hand and removed the charcoal which he then showed me, somewhat shamefacedly. It was still quite damp: after removing a caruwa you should blow on it in your hands for a few minutes, which dries it off.

Chico knew that he hadn't put any charcoal inside me, but he'd put something in, a bit of himself; which, since I didn't take it as seriously as I should have, was, I suppose he felt, left at my mercy. He was getting back the power of his own imagination, which had over-reached itself. Indians know that the taboos they obey often have no effect on the white men who break them, but they feel that the magic involved may rebound and hit one of themselves, though they are merely bystanders. Magic, once set in motion, is not particular where it has its effect.

There was a lot of talk about shamans the next morning. Chico said that the king-vulture had sent him a caruwa, which made him cough and cough so that he nearly fell over: the king-vulture's caruwa is a maggot. Someone had to fumigate him with tobacco, to get rid of it. Taperé told how a shaman had come to his village, sung and smoked his cigar, and all of a sudden he looked up at the roof of the hut and money started falling down. 'Pick it up!' he said; and people went to pick it up but couldn't find it. Then he smoked a bit more, reached with his left hand under his right shoulder and produced a stone, the size of an almond: and he rubbed it between his hands, rubbed it and rubbed it, and when he opened his hands it had gone.

That, of course, was one of Chico's tricks too. But there was one trick Chico could not manage, that of killing a man at a distance by magic. To do this, the shaman 'softens' a specially made wooden caruwa and blows it, with a puff of spiritually charged breath, towards the man; no matter how far away he is, the caruwa will find the man, enter his body and kill him. Chico was unable to perform this sorcerer's trick because he had once touched a corpse, and that one corrupting contact had bled him of his power. This is one of the paradoxes that are so common among the Indians. When a shaman, or a sorcerer, falls into a trance he is said to be 'manon-manon', dead-dead, and only then is he able to converse with the spirits that come to inhabit him; and his ability to do this confers on him a kind of immortality, since after every such 'death' he wakes up, or is 'born' again. To confuse this spiritual death with a physical one, however, is fatal, and if he should touch a dead body his magic is short-circuited, leaving him empty of power.

In order to keep his power intact, the shaman has to lead a somewhat secluded life. He is greatly admired, for if the power to do things is a kind of commonplace magic, the power to make things do themselves is irresistible: Indians take great delight in a watch, or a magnet, or any more or less self-moving object. But the shaman's power, which he derives from a dangerous and laborious mastery of the spirits, must be used productively if he is not to turn into a sorcerer, and his power into black magic. A good shaman has tricks, but his reputation rests on the cures he effects: that is, he may not keep his power to himself. If he does, and uses it selfishly, he may be put to death: a fate which has overtaken several Tembé shamans in the past, and is in fact the occupational hazard of shamans throughout South America.

Without personal power, however, no one can be a shaman. Antonio-hu would call me *payé* in joke sometimes, for he had once slept in my hammock and woken up with a headache due, he said, to a caruwa I had left behind; and once he heard me saying 'tut tut' when my elbow was jogged, just as a shaman would have done. But I could not be a real shaman, despite the medicines I had brought with me, because I had no familiar spirit, and because the songs I knew, which the Indians were always asking to hear, didn't cause anything to happen. I never even used a cigar when singing, which proved I was no shaman: for tobacco smoke, being not only a drug but a fumigant, is as it were the spiritual breath made visible, though it only has meaning to the Indian when he can make it his own, and add his will to its power. It was for this reason that Chico forbade Tero to eat anything that had a smell: the smell of roasted peccary is a kind of spirit, and interferes with the refined spiritual action of the caruwa.

The magical powers of the shaman, and his 'immortality', show him to possess many of the characteristics of Maír. Now Maír can do anything: turn unwanted visitors into stone, bring the dead back to life, create animals, and produce manufactured goods. He can even make ordinary people immortal. The Urubus tell of a couple whose child had died, and who decided to visit Maír and ask him to bring it back to life. They walked to the sea, and boldly plunged in and swam in the direction of the rising sun; in two days they reached the island where Maír lived. Maír greeted them affably, and was told the reason for their coming; so he gave the man a headdress, and washed both him and his wife in a magic water. Then he made them catch hold of a low branch, so that they hung from it with their feet off the ground, pulled at their legs to lengthen them, kneaded their bodies to make them strong, and gave them each ten shirts to wear. Five were soft, like cotton, and five were hard, like stone; when the couple were dressed, Maír pulled out his machete and gave them a few hacks, but the machete just bounced off; then he took his bow and arrows and shot at them, and the arrows bounced off as well. "Now you're like me," said Maír. But as the couple were still angry over the death of their child, he took its bones, which they'd brought with them, blew tobacco smoke over them and over the woman's belly, then put the bones inside her. In four days her belly was sticking well out and on the fifth day she gave birth to her son all over again.

Immortality is really only another word for rejuvenescence. The fact that Maír gave a number of shirts to the couple points to the fact that he lives for ever because he can always slough off his old skin, like a snake; and that the shirts were hard, like stone, shows that he was giving them what Buddhists would call a Vajra-body, a diamond body of immortality. That the couple were also lengthened fascinated me, for it explained why the Indians suspected I might possibly be immortal: I am tall, as though I had gone to visit Maír and been lengthened. Indeed, many of the Indians asked me to pull at their arms and legs, to make them taller and so, presumably, give them long life; and I had one trick that was always in demand, that of pulling my arm out of my sleeve in such a way that it seemed to be growing. But Chico, when he blew a strengthening caruwa into Tero and the other men, and then kneaded their bodies to 'arrange' it, was also imitating Maír: alas, an inferior imitation, since Tero was able to be immortal for only three days.

CHAPTER SIXTEEN

Rain

THE rain, which had been light, now began to fall heavily once or twice a day. One could not forget it: even when the sky cleared a little, the jungle could be heard roaring slightly with the drip of water falling from one leaf on to another. The jungle looked different as well, as I found when I made a trip to Canindé to get the rest of my stores, brought upriver by Feliciano on one of his trips. There were fungi growing by the path—I remember especially a very large *Phallus impudicus*, looking extraordinarily grotesque with a kind of umbrella in white and bridal lace growing just below its head. Ripe fruit had dropped in great numbers from the trees; the streams, dry all summer, were now full of water, and small fish and terrapins were to be found in them. Where the ground was swampy, among the roots of prickly palms, red and black land-crabs reared back on their legs and threatened us with their claws. The armadillo I had shot was now nothing but bones, and its carapace looked like a spilt jigsaw puzzle; and several score more rhinoceros beetles had come out of their burrows to die in the open.

The birds were nesting; pigeons in low bushes, toucans and parrots in the holes of trees. The path was criss-crossed with the tracks of tapir, and peccary, and jaguar; and from time to time the bushes would shake convulsively as a startled deer fled away from us. With so many things to see, it was a pleasant trip for everyone, except Mbeiju who was feeling ill, and who wept several times because, she said, the stones hurt her feet, and she couldn't stop falling down, and her load was too heavy

—though she carried nothing but a wicker back-basket containing two hammocks, and three puppies curled up on top.

We came back some days later, having left Picher behind. He had caught *curuba*, the jungle itch; and, being cook, had already passed it on to Chico and myself. Curuba is an intolerable malady, for one can hardly stop scratching oneself, and Picher had got it so badly that his thighs were quite raw, and he could hardly walk. Sulphur clears it up, but I had none; so Chico and I tried a cure proposed by the Indians, the juice from a certain leaf and, when that didn't work, we scratched ourselves till we bled and daubed our wounds with lemon juice and gunpowder. It was a painful but effective remedy.

The rains increased; great storms darkened the jungle, with much thunder and lightning, and when the wind blew one could hear the lengthy crash of falling trees, their roots loosened by the continual rainfall. The night after we returned to the village, indeed, a large wild cashew tree suddenly fell down behind Toí's hut—'aha!' we all cried, 'aha!' and Toí, fearful that other trees might fall on to his hut, spent the rest of the night with us.

In the next week, we cut down a number of trees: some for their fruit, on which we gorged, and some to get at birds' nests. We found a dwarf parrot's nest, with five squabs which Mbeiju took home and fed with manioc pap out of her mouth; and some toucan nests, though the birds had already flown. One of the trees we cut down was a giant, falling with a roar of snapping branches and lianas: it made a great hole in the woods, for it brought seven other trees down with it, and lay in a clearing of its own making, sprigged over with epiphytes on trunk and branch and twig, a hanging garden. Its core was quite rotten, full of a strange red clay made of dead wood, dust, leaves, and other things, and there was a root in it from a plant that had established itself in a crack forty feet higher up.

Pari and Chico found a bees' nest in the tree. They were small black bees without stings, but very angry: they mobbed Pari, getting tangled in his hair and crawling into his ears and his eyes. "Aha!" cried Pari, "hahh! hah-hah!" as the bees tormented him.

"Don't shout!" said Chico sternly, "it's bad for the honey, it won't be there if you shout!" But Pari went on shouting, he said he always shouted when cutting out a bees' nest; he only wondered if he shouldn't have whittled some bits of bark into the shapes of knives and machetes, so that he could leave them by the side of the nest in exchange for the honey.

"If you take honey from shaman bees without paying," I was told—shaman bees are a particular species, but all bees have shaman powers—"then you quickly fall ill, you get awful pains, you vomit blood, you're dead. And if you're married you don't take honey. The shaman bee knows, he says to himself, 'That man's married, he's got a pretty wife; I want her!' So he takes her and she falls ill and dies. A young man, that's all right, if he's not married he can take the honey; he cuts the tree open, pulls the honey out, then he puts into the hole a machete, a knife, a mirror, a pair of scissors. He makes them out of bark and puts them in the hole. 'Ah,' says the bee, 'that man's good, he knows, he pays when he takes honey! A machete, a mirror, a knife, he pays properly!'"

Pari didn't bother about making bark knives, he just pulled the combs out and squeezed the honey into a large gourd bowl. It was a thin clear honey, very slightly sour from the pollen which had got into it, but extremely good. Chico saved the combs in which the young grubs developed, dipped them in the honey and ate them. I did too, seeing Chico's look of satisfaction: the grubs tasted of very young hazel nuts, a milky innocent taste which

I thought the honey spoilt. We had them all to ourselves.

Antonio-hu told me later that in the old days they used to have a honey feast, but no one else knew anything about it. Perhaps he'd heard about it from the Tembés, where it's an important festival. They'd hang the honey up in pots, Antonio-hu said, under the rafters, and the women would sing underneath, all through the night, for several nights, while the men sat round the sides of the hut. The feast was called Irar, after a small black animal with a white face (a species of grison) that eats honey. The women would sing the Irar song, singing that Irar sits by the bees' nest putting its hand into the honey and licking it, slup slup slup!

After several nights of singing the honey would be taken down, mixed with water and made into a shibé. "Ah," said Antonio-hu, "when we drink the honey we don't stop till it's all gone, we get full bellies, just with honey. And then we suruk our wives, then we make children, our rankuaïs are all sweet. Sweet sweet sweet."

I was surprised, considering the season, that there were so few mosquitoes about, since there were lots of pools in the jungle for them to breed in. The year before, Pari said, there had been so many mosquitoes that he couldn't sleep at all, so he took his hammock into the woods and slung it between two trees. But the mosquitoes found him out every time, though he moved all round the village trying to escape them. Even a good smoky fire underneath the hammock didn't stop them: balked in their favourite direction of attack from underneath, they came at him from the top.

There are some years with plagues of mosquitoes, other years with hardly any. It was luckily one of the scarce years, and I was never troubled during the nights. The cicadas came out, however, from their burrows in the earth. Mbeiju took several home and watched with Pari as they struggled out of their larval skins. They would bulge out slowly, slightly crumpled but magically coloured in red and green, with the wings small buds of vermilion, till in a few hours their wings had expanded and their skins hardened. "Just like Maír," Pari would exclaim as he watched: Maír too sheds his skin.

The rains filled up the small dry creeks near the village, and brought fish up out of the river to spawn. Toí went with his family one day to fish a pool with poison. He cut a special kind of liana into lengths, tied them into bundles and beat them in the water with a heavy stick, so that the frothy sap came out and spread through the pool. The sap affects fishes' breathing, so that they dash wildly about near the surface, gulping in air, and they get slower and slower till they can be picked off with an arrow, or sliced with a machete.

One of the girls came home crying: she'd put her hand into a hole in the bank, after a fish that had got in there, and had been bitten on the finger by a snake. It was luckily not a poisonous one, as Paijé found out after a brief inquisition, not like the one, she told me, that had bitten her on the head years ago, and had nearly killed her. Then too she had been out beating fish-poison in a pool. She bent down under a bush and felt what she thought was an arrow pierce her scalp—"Don't shoot at me!" she yelled at her husband. "But I haven't even seen a fish to shoot at!" her husband replied: so then she found it was a snake that had bitten her, and the blood was running into her eyes from the wound. Her head swelled up as big as a water-melon, and Anakãpuku and Pari both tried to suck out the venom—they sucked and spat and sucked—and Terepik, who was pregnant at the time, tried to help but was chased away because if a pregnant woman sees someone who's been bitten by a snake her hair falls out and then her flesh falls off and she dies.

I hadn't been back from Canindé many days when I realised that Pari was getting a bore, and that the village, with its dirty and unkempt look, depressed me. So I got most of my things together and went off to the neighbouring village, where Antonio-hu was expecting me. I had already spent a few days there, and had enjoyed myself a great deal: the village, though small, was neat and handsome, with a prosperous air about it, the clearing being full of flourishing manioc plants, seven feet high.

Antonio-hu had built a new hut against my coming. He told me that he was ashamed of his old one, ashamed that a caraí, a white man, should have to sleep in it, under a leaky thatch; so he built a new one, such as I've not seen in any other village—forty or fifty feet long, and eighteen feet wide, the top of the roof some twenty feet high. With great care he had built a table out of cut saplings for me, and a bench to go with it, where I could write: he laughed with joy when he saw how pleased I was with it.

But the village was nearly empty when we arrived: everyone had gone hunting except for a middle-aged man called Wirakangupik, who was convalescing after a bout of fever, and his wife Mirijun. Wirakangupik was the most ingratiating Indian I ever met: he had a kind of false party smile that rarely left his face, and never lost an opportunity to flatter me. "Eh, chief Arajuba!" he would call if I passed: the arajuba being a rare, completely yellow parrot much prized by the Indians for its beauty. I would retaliate by calling him chief Agouti, that rodent famous for its sexual powers; for he was the father of Saracaca, and in his youth had been as intemperate as she was. His father, interestingly enough, had been a Guajaja caught by the Urubus as a boy, and given two wives by them. I do not know if he also was a lecher, but he loved working, and passed that redeeming trait on to his son. In fact, Wirakangupik worked more than Antonio-hu did, though Antonio-hu was the chief; he and his wife more or less ran the economy of the village, getting the girls to dig up manioc and roast it when it was needed, making shibés and messes of tapioca.

Wirakangupik greeted us with his appalling smile, and gave us a shibé to drink. It began to rain again, violently: the ground flooded, and spray came drifting into the hut in misty clouds. We dug a shallow gutter around the hut below the eaves, to drain the water off, and waited for Antonio-hu and the others to come back.

They had been unlucky, killed no game, seen no tortoises, gathered no fruit. They stalked disconsolately through the rain and stood round the fire, planting their bows and arrows into the earth to dry. "I want a cigarette," Antonio-hu said brusquely, and soon I was making cigarettes for all of them. "I want sweet manioc soup," I then retorted, and Antonio-hu hurriedly set the girls to digging up a huge tuber of sweet manioc from a patch just near the hut, which was peeled and boiled down into a thin soup, to which I added some of my rice.

We sat round together, drinking it out of gourd bowls. Saracaca looked at Chico with hope in her eye; and so did her husband, lean, intellectual Alishandre, though for very different reasons: he was learning to be a shaman, like Toí, and hoped Chico would help him. Saracaca's brother was also there, with Curumiú, his bead-stealing wife, and several children, mostly girls.

Alishandre came and sat next to me in my hammock. That very morning, he said, the birds had been singing, singing so loudly that everyone thought, "It must be because Francis is coming today and bringing us beads, that's why the birds are so happy!" So I took my tin of beads out of my rucksack and shared it out. I liked giving things in Antonio-hu's village: I could see how much they liked what I gave, and they never pestered me for more, unlike Paijé who could not be satisfied.

"You're my friend," Antonio-hu said suddenly, "when you go home again you must tell everyone that we're good, tell them to come and visit us and bring us cloth, and beads, and knives, like you've done. Tomorrow I'll go hunting again, I'll kill a deer for you—that's where I'll kill it," he said, poking me in the ribs with a finger.

It stopped raining; the girls hurried into the clearing to get some more firewood, and Antonio-hu, Chico and I went down to the stream to bathe. Antonio-hu took some flour in a bowl, to make a shibé; and after he had prepared it he borrowed a tablet of soap—a scented one, my one luxury—and soaped himself white, all over. "What a good smell!" he said.

The moon came up. It was a lovely night, without a wind, the tree-trunks at the edge of the clearing standing out remarkably white; and a lively noise of frogs down by the stream.

We drank some more sweet manioc soup, rolled ourselves large cigars of tobacco and tawari bark, and then Antonio-hu and I lay down at either end of my hammock, head to foot. One lies diagonally in a hammock so as not to be bent up like a banana, and by oneself nothing could be more comfortable. Two in a hammock is companionable, but a bit squashed; still, the Indians like it, especially the young men. I've seen five in a hammock before the cord broke.

I asked him about the stars. He told me about the Milky Way, which he called the Tapir's Path: the souls of dead tapirs walk along it, eating 'fruit souls and leaf souls,' as he said. There was Grandfather Many Things, or the Pleiades, each of which was really a man dressed in his feather ornaments. Antonio-hu pointed to the east—"When Grandfather Many Things shows itself there," he said in his deep, grating voice, "it's summer." He lifted his pointing finger to the zenith: "There, the rains start! When Grandfather Many Things is there, we know it's time to plant." His finger pointed slowly to the western horizon: "The rains are over."

The Pleiades are an important constellation, which is why they are called Grandfather: the only other constellation so honoured is the Southern Cross, or Grandfather Cotton, though it does not seem to have any connection with agriculture. The Morning star is also important, being called the Mother of the Many Things, as well as the Moon's Sister; but nearly all the other stars are thought to be the eyes of animals and birds. The only two exceptions I noted were Mars, known as Macaw Egg, and Sirius, called Nenguru, a name whose etymology I did not understand.

I asked, idly enough, if they did anything when particular stars came up over the horizon for the first time every year. It was a good question. In the old days, when Macaw Eye, or Aldebaran, first showed itself, the Indians would take macaw feathers, burn them and smear the cinders on their children's foreheads; then they would take a fish-tooth and scratch their arms so that they would grow up to be strong, and good hunters of macaws. These two results mean the same thing, for the macaw is a symbol of the sun, as will be seen in the next chapter, and to shoot macaws is a symbolical way of being a hero. Then there was Deer's Eye—it was still below the horizon at the time, so that I could not identify it—that, said Alishandre, was a big star. He made a fist: "bright!" he said.

"What do you do when you see it?"

For answer, Alishandre seized hold of a young boy who was near him, held his arms and then made as though he were drawing a bow. "You *shoot* at it, you make the children aim at it, right in its eye. With a takwara, an iron-headed arrow."

"Why?"

"So they'll always kill deer," he said, reproachfully.

And he told me about Three Eyes, Orion's belt, that

when it appears children wake up early in the morning when it's still dark and go and bathe in the stream to become strong; and how stars are attached to the sky by a little stalk, and when the stalk breaks they fall and become shooting stars. He had come across a shooting star once in the jungle, lying on the ground; a soft jelly-like object,* he said, shining blue, and very hot.

Chico was flabbergasted. "And I had to come here to find out about all these things," he said, "all this about the stars." He was especially put out because the Tembés look down on the Urubus as being ignorant and coarse.

"You don't ask, you don't find out," I said, rubbing it in.

"That's right. No one ever asks my father so he never talks about things like that. Eh, shooting star!"

Antonio-hu rounded off the evening by telling us two long and rather boring stories about raids, full of flying arrows and slaughter; but he told them with such conviction, and in such detail, that I had good hopes of getting him to tell me a Cannibal Tale before I left. We finally got to sleep some time after midnight. I woke up shivering at four, and heard Saracaca blowing the fire up beneath Alishandre's hammock; at daybreak it was raining again.

* Probably the jelly-like alga *Nostoc*, frequently—as in England—called 'star-jelly'.

CHAPTER SEVENTEEN

Myth and the World

ANTONIO-HU, like all the other Indians, was fascinated by aeroplanes. We could hear them occasionally humming in the distance, and from time to time one would come overhead. Antonio-hu would go outside and look up dreamily. "Tell it to come down," he would say; and after it had gone he'd ask if one could sleep in an aeroplane, have meals there, and if it wasn't terribly hot so close to the sun. "Prrrrrrrr!" he'd say, and make a large gesture with his hand into the sky.

"What happens to the sky?" he asked.

"It goes on," I said.

"Doesn't it come to an end anywhere?"

"It goes on and on, it has no end."

Antonio-hu shook his head slowly. He thought the sky was hard, and that it started at *iwi pita*, the place where the land stopped, the end of the world.

"Have you seen iwi pita?" he asked anxiously.

"He means the sea," Chico put in, though Antonio-hu didn't mean anything of the kind. I got out of my hammock and started drawing a map on the ground. Brazil was under Antonio-hu's feet, North America was below our rustic table, a dog had to be shooed away to make room for Europe, Africa was by the waterpot, and China was on that strip of earth under the thatch where the rain gutters. Australia had to be drawn round a tree-stump, and New Zealand was very nearly in a manioc bush.

Antonio-hu was not convinced by this diagram of the world. There were too many countries in it, with too many names, and though he liked hearing about them he was troubled that I could not point iwi pita out to him. Eventually he burst out, scornful of my knowledge and of the boasts of caraí, of white men—"Caraí tell me they've been to the end of the world but they haven't, they've just been from here to Belém. The end of the world is where Ara-yar (macaw-owner) lives. I ask caraí, 'Have you seen Ara-yar?' 'No,' they said, 'no one's seen him.' "

One man, he told me, had seen Ara-yar: it was yandé ramuĩ again, our forefather, who had found himself by chance at iwi pita, near a village full of macaws, yapus, parrots, parrakeets and other birds. It was a strange country, because everything was burnt: the trees were charred stumps, and the ground black with ash. When he was still some distance away from the village yandé ramuĩ could hear Ara-yar pounding mindubi fruit to give to the macaws, but when he was in the middle of it all he could see was a small palm-leaf mat with the pounded fruit lying on top of it; there was no one in any of the huts. When he left, the pounding started again.

"No one knows who it is," Antonio-hu said dolefully.

"It's Maír," I said, putting two and two together.

"Maír, of course, he's Maír! Maité," he said in admiration. "Have you seen Maír?"

"No."

Antonio-hu sniggered. "When caraí go to see Maír, he turns them into stones to sharpen knives on."

"That's right. And when they paddle up the river there, they paddle and paddle but they can't get anywhere, the water's like rubber," I said.

"You know! You're right! They paddle, they can't get anywhere; they get on to the bank, their feet stick there, they die from the sun."

"Pirambir saw Maír," he said suddenly. Pirambir was a yandé ramuĩ whose name he remembered; he'd seen Maír, and reported that he was a beautiful man who wore a headdress, a feather collar round his neck, and other feathers round his arms and his waist. "Pirambir said Maír was a real caá-por, a real Indian, that he was living over there, where the sky comes down to the earth."

"The sky meets the earth?"

"That's right. That's iwi pita, where Maír lives. The sun comes out close to his house, climbs up the sky——" Antonio-hu traced its course with his stretched arm—"and goes down there."

"How does it get back to Maír's house?"

"The sun's a man, he's like a man; when the sun sets, then he walks back again underneath us. There are woods, rivers there, it's a world just like this one."

"Who made him?"

"I don't know. Maír made him! Made him all beautiful, and put a headdress on his head."

Antonio-hu leant back in his hammock and looked at me. "You haven't heard from other caraí about the time the sky came down on top of the earth?"

"Speak," I said ceremoniously.

"The sky came right down, it was only as high as the top of this hut. Black! no one could see, black, black! Everyone waited in their hammocks: one day, no sun: another day, no sun; another day, another day, another day, no sun. Everyone was hungry. They began to cut through the sky, they cut and cut"—Antonio-hu took up one of my aluminium plates and hit it with the side of his hand—"it was hard! They cut through it, made a small hole and then five people went through the hole into the sky. Then the sky went up, right up, to where it is now. The others all stayed on the earth, they didn't get through the hole; if they had, we should all be up there now. They couldn't find the hole, it was too far up."

"When my wife was dying," he continued, "she told me, 'Don't cry, don't be sad. I'm going up into the sky, there I won't die any more. When you die, you'll come up too, we'll be together again. You'll see, the sky comes right down to you, it's only a little way to go—you just step in and you're in the sky. Don't be sad!' But," said Antonio-hu, "I was sad, very sad."

It is easy to say that this picture of the world, which Antonio-hu described, is a mythological one. It is: but, more to the point, to the Indians the picture is a real one. They know the world as a flat place, with the sea at one edge, or perhaps all around it; the sky is a hard dome which rests firmly upon the earth. Beneath there is an underworld which allows the sun to pass at night from the west to the east. The three worlds, in the sky, on earth, and underground, each have their distinctive inhabitants. The earth is, naturally, the world of the living, while the sky is where the liberated souls of the dead live. The underworld is for ghosts, the mortal emanation of dead bodies. When a man sees the sky coming down upon him, he knows his death is near, not only because he can so easily step into the world of the dead, but because the universe is reverting to a state where no one had yet said, Let there be light.

It is rare for primitives to think about a creation from nothing. For the Urubus, the beginning of the world is marked by the emergence of Maír from the jatoba tree, at the same time as Vulture (from whom Maír stole fire) came out of the pau d'arco, and man out of the wira pitang or red-wood tree. Maír, they say, often burnt the world, put the fire out with a flood, and then replanted it: but everything was there before he appeared.

The Indians are interested in another kind of creation, like that told in Greek mythology where Night, impregnated by the Wind, conceives and lays an egg which later splits in two. The top half becomes the sky, the bottom half the earth; out of the egg comes Eros—or, as he is sometimes called, Phanes, he who shows. He is in one sense the equivalent of light or consciousness.

The Indians have a less abstract imagination, and they

see the separation of the earth from the sky as a temporary event, personal to every human being. When a man is born, the sky rises; when he dies, it descends again. In between, the sun appears.

The sun, like so many other things in the Indians' world, is not only what it is seen to be but also symbolises some human virtue: it is also a man. In spite of Antonio-hu's remark, no one really knows who made the sun—Maír is always mentioned as creator when people don't quite know how things came into being. Things which were definitely created have stories attached to them, explaining how they came to be as they are: things that always were have no stories. But there is a good reason why Indians do not know who made the sun, since the sun is Maír himself. That is why iwi pita, where Maír and Ara-yar live, is burnt up: it is the sun's place. (Maír is also known as Ara-yar, and this seems to point to a meeting of two separate mythologies. But the macaw (Ara-yar means macaw-owner) is a good emblem for the sun, for various reasons. Its feathers are bright red, like fire: it has a huge, powerful, curved beak, symbolising fierceness, and it commonly sits at the very tops of trees, as near to the sun as possible.)

Ara-yar's village at iwi pita must be somewhere on the edge of the continent—if one was looking for a real place one would think of the north-east coast of Brazil, where droughts and floods are a tragic commonplace. This would make factual sense of Maír's repeatedly burning and flooding the world. Iwi pita therefore differs from Maír's mythological island, which the Indian couple (Chapter 15) reached by swimming. This island is an Isle of the Blest, such as exist in other mythologies, where the heroic dead live out their care-free after-lives; while Ara-yar's village is the sun's home, where living men are an anomaly. But Maír's house on the island is also a sun house, since its walls and roof are said to shine brilliantly;

a sun house not for the real sun, but for the spiritual one.

All Indians want to see Maír, though they say that with the coming of white men he has moved away from his old home and can no longer be found. Their favourite story concerns his children at the beginning of the world, and how they made the great journey to find him and prove themselves to be his sons. It begins with the episode of Maír's setting fire to the world, and then putting the flames out with a flood, after which he replanted the jungle with smaller trees. He did this several times, and the Indians complained to him about this, because each time they had to climb trees and might be turned into sloths, if they screwed their eyes up against the smoke, or into frogs, if they slid down the trees for a drink before the flood had gone. So Maír turned temporarily into a yapu bird, and flew away. He flew past an apu-i tree and saw one of its fruit, full of worms (as apu-i fruit usually are) lying on the ground. 'That would make a nice woman!' he thought. Sure enough, later that afternoon a woman walked up to him when he was having a rest, and he made her his wife.

That afternoon he cleared a bit of jungle, planted manioc, and cara potatoes, and water-melons; that night he slept with his wife. In the morning he told her to get him a water-melon from the garden, but she said he'd only just planted the melon, how could it have produced a fruit? This made Maír angry, and he left her.

After a little she followed him, guided on her way by the child Maír had made in her belly. She stayed the night at the hut of Mikur the opossum, who seduced her and made a second child in her belly. In the morning Maír mimi (mimi = child) would no longer speak to her, and sulked in her womb, so that she lost her way and came to a village of jaguar-people. Everyone was out hunting, except for an old woman who hid her underneath a large pot. This didn't stop the returning hunters from finding her out, killing her and eating her; and as they ripped her belly open, the two children fell out. The jaguar-people wanted to eat them too, but each time they threw them into boiling water the children jumped out again, so that they were given to the old woman to look after.

The twins grew up very quickly: in three days they were old enough to go hunting. There in the woods they made themselves axes and machetes by magic, and at the edge of a stream they cut trees into the water. The trees turned into caymans and piranha. Then the twins made an island in the stream, and they turned the stream into an enormous piece of water. On the island they made naja and piqui trees, bearing enormous fruit; and they connected the island with the land by means of two liana ropes.

When they returned to the village, they took a lot of naja and piqui fruit with them which the jaguars liked so much they insisted on going to the island to get more. Next morning the twins led them there, and told them to cross the water by means of the ropes, which they cut when all the jaguars were upon them. All but two fell in the water and were eaten by the caymans and piranhas; and if it hadn't been for Mikur mimi's clumsiness, none would have escaped.

The twins then went to find Maír. Maír was making two bows, one for each of them, but the twins were angry with him for having let their mother die, so they picked up the bows and threw them into the jungle, where they turned into red snakes. Maír ran away from them, but soon returned and proposed a test to see whether they were truly his sons. They had to go to the place where Anyang the ghost did his fishing, and steal his bait.

Anyang was fishing in a deep pool, and Maír mimi at once jumped in, removed the bait from the hook, and gave the line a few tugs to tease Anyang, who struck and missed. Then it was Mikur mimi's turn and, though he'd been told what to do, he put the hook into his mouth, after he'd taken the bait off, and then tugged the line: Anyang struck, and Mikur mimi was caught.

Mikur mimi was eaten that evening, but his brother turned into a large tapi-ai-ĩ ant * and collected all the bones, breathed on them and brought Mikur mimi back to life. They returned to Maír who proposed another test; they should go and see Iwitu ramuĩ, Grandfather Wind. Iwitu ramuĩ was out when they came, but his wife was there, and Maír mimi began to tickle her so that she laughed. She laughed so much that her husband heard, and got very angry—'Who's that making love to my wife?' he shouted, and began to blow a gale. The twins were scared and ran away, but Iwitu ramuĩ went on blowing till he managed to make a tree fall on top of them. Mikur mimi again died, but his brother again brought him back to life.

Maír then set them a third test, which was to pass through *ita kitik*, the rubbing stone—sometimes called *ita kuí*, the grinding stone. It stood in front of his house, opening and shutting. As it opened, Maír mimi went through, and when it shut he was safely on the other side: as it opened again, he returned. Mikur mimi tried to do the same, but he was too slow coming back: the stone closed upon him, and there was nothing left of him but a spot of blood. Maír was pleased, because he'd always thought Miku mimi, who was yellow, was not his son; Maír himself is black. Maír mimi, however, was very angry. He breathed on the spot of blood, and his brother came back to life once more.

Maír was disgruntled at this, and made Mikur mimi go through ita kitik yet again. Mikur mimi was still too slow, and again ita kitik closed upon him; this time there was nothing left behind. Maír mimi was so angry that he

* The ant used in girls' puberty rites, and in the rites following an execution.

took ita kitik and went up into the sky with it. Up there he began to shake the world, and repeatedly thrust down the sky: Maír then had to push it up again. But the first time it happened, Maír turned into a crocodile, and went to hide in a pond. Maír mimi made the world shake a little more, and began to rub ita kitik together: that caused thunder, and that's how thunder is made now. But the story does not end unhappily, with Maír mimi still in a rage, for Mikur mimi went up to the sky when he died, and met his brother there.

It is worth picking this myth to pieces, to see what it is made of. It starts off with the Indians complaining so bitterly against Maír for burning and flooding the jungle that he moves away in disgust. In Tupi myth Maír is chased and shot at, and sometimes he even turns into a stone, the typical fate of the unsuccessful hero. This hero exaggerates everything, even his own masculinity: if it wasn't for women's incredulity and laziness, the Indians say, crops even nowadays would bear the day after they were planted. The same point is made in the story of the rankuaĩ-ang (Chapter 12), where Maír could make children by magic only when women didn't pry and interfere.

This is the Fall: blamed on women, for not sharing man's beliefs, but properly the fault of men, for romancing. After the Fall, the exile: acted out by Maír who stalks grumpily away, leaving his woman behind who perforce has to follow him. She has already conceived by him, his magical pretensions notwithstanding, and his child immediately shows its magical nature by speaking out of her womb, and guiding her in his father's footsteps. The woman, untrustworthy as ever, lets herself be seduced by Mikur the opossum, and conceives another child: and Maír mimi, affronted by having to share the womb with an intruder, sulks just like his father and offers the woman no more advice.

The opossum is chosen as a seducer for two reasons.

The first is that the male opossum has a forked penis, just as the female has a forked vagina: he is thus an excellent begetter of twins. (In other parts of the world the opossum is thought to copulate with the female in her nostrils, since her double vagina has not been noticed; she is believed to give birth to her young by sneezing them out.) The second reason is that the opossum shams dead, in fact plays 'possum'. This displays the character of Mikur mimi, the opossum's son, the mortal twin to immortal Maír mimi, and explains paradoxically why the opossum is thought to be a shaman: for the shaman can only be possessed by his familiar spirit after he has fallen manon-manon, dead-dead. The ability to sham dead also explains why the opossum is often equated with the moon in related mythologies, as against the Maír mimi figure of the sun, since the moon is notoriously mortal and changeable: characteristics developed when the opossum is turned into a trickster hero, holding his own by wiles and evasions against brute force.

Big with her two unborn children, the woman blunders into a village of jaguars. This village, like the surrounding country the woman has passed through, is underground, the land of the night sun: the jaguars are thus to be seen as the black side of Maír's nature, for one of Maír's epithets is Yawar-iyar, Jaguar-owner.

What happens then is an inevitable outcome of male pretensions and savagery: the woman is discovered, killed, her belly ripped open and an attempt made to destroy the twins. This is a mark that the twins are powerful spiritual forces, since their birth antagonises what is dark in the world: and it can also be read as a dramatised description of ordinary childbirth. The old woman who hides Maír's wife is called Aí, Grandmother, and is in fact the woman's mother helping at the birth: the twins' adventure in the cooking pot shows them being washed in warm water.

The twins, being magic, grow rapidly, make iron tools out of bark as Maír does, and create a large island in the middle of a lake. This island is similar to the one on which Maír lives: it is an isle of the blest. They plant it with piqui and naja trees, whose fruit are the favourite food of the opossum, and entice the jaguars to visit it. The jaguars, being bestial, cannot reach the island which is for the spiritual alone, but are cast into the water and eaten by fish. Two of them escape, owing to Mikur mimi's stupidity, the stupidity which is the other face of his changeability. The old grandmother, when they return to the village, is changed into a tamandua anteater, the Curupir man of the woods; the twins have now proved their spiritual nature up to the hilt, in addition to revenging their mother.

They go to find their father. He is making bows for them, a tacit avowal that he is their father: for men not only make bows and arrows for their sons, but present them with miniature ones, together with tawari cigars, fire-sticks and macaw feathers, when they are born. The twins turn the bows into snakes, which frightens Maír so much that he runs away: it is clear they are his equals. But he sets them three tests.

The first is to trick Anyang. Anyang is a corpse, everything that Maír is not; and the twins when entering his fishing pool are like Jonah being swallowed by the whale. It is almost a tautology, for their whole story is a Jonah story, a descent into the underworld: and the particular form it takes may be connected with an old Tupi custom, a father fishing for the carrying sling of his new-born son, with a hook and line. This was presumably to make him a good fisherman, but magic, like the proverbial stick, has two ends to it, and bad Anyang lies at the other end of good fisherman. Mikur mimi swallows the bad end, and dies, true to his nature; his brother resuscitates him.

The second test is to confront Iwitu ramuĩ. This looks like a mythological account of a boy's first sexual adventure, usually with a woman much older than himself: it is told in another myth, where a boy rapes an old grandmother anteater after killing her husband, who had previously tried to kill him with a shattering fart.

The third test is to pass through ita kitik. Here we are back at that prime mythological virtue, hardness. Ita kitik itself seems to be based on a grindstone, and in that form probably comes from Peru or the Caribbean, where grindstones are common; but the idea of clashing rocks or Symplegades is universal, and symbolises the brutal self-contradiction of the terrestrial world which the hero must pass through before he can enter that other world in the sky. Ita kitik is a rock, and hard (even though one term of the contradiction is the soft corpse of Anyang), because the hero in confronting the world must harden himself: if he tries, and fails, he does but grind himself to death. The failed hero, in fact, is petrified, and turns into a stone, instead of acquiring a living hardness.

Mikur mimi, being only the mortal part of every man, is destroyed by ita kitik. Every voyager into the underworld leaves a part of himself behind, if it is only his tail —the Argo, it will be remembered, left its stern ornament behind in passing through the original Symplegades, when sailing for the Golden Fleece. Theseus abandoned Peirithous in the Labyrinth; King Arthur left Gwair in Caer Sidi; Gilgamesh lost Enkidu. Mikur mimi's death gives Maír mimi his triumph: for Maír mimi has not only shown himself immortally hard, but vanquishes his crocodile father and carries ita kitik off with him into the sky, where he now demonstrates himself by thundering victoriously. He has now become what his father was. Mythological figures have no more power than the attributes they possess: steal their attributes, and you have their power.

The end of the myth is inconclusive. Maír mimi has

gone to live in the sky, where he receives the spirits of the dead. He himself, of course, is spirit, opposed to Mikur mimi's body: and he is therefore in one sense dead himself. He is the heroic part of man, but not the living hero himself: that is left for Maír to embody. Maír's transformation into a crocodile is a temporary affair, for he soon reverts to his human shape: he is in fact the type of every man with chief-like aspirations, and to make this plain he lives not in the sky but on the earth, and can be visited by such living men who are hardy enough to do so. Maír's power, therefore, can never be taken away from him: the man who defeats him is a living man also, and thus becomes a Maír in his turn: he only becomes a Maír mimi in the other world.

The myth of the twins in search of Maír is common to so many tribes in South America that it must be seen as a general account of the mythological world, and not a particular one. The only telling variations are those which leave Mikur out of the story completely, with the consequence that his son does not figure in it either; or which make Mikur the equal of Maír. The story can thus either be about equal twins, unequal twins, or a single man, confronting the figure of their father; and these variations reflect differences in the workings of society. Among tribes who practise cross cousin and avuncular marriage, a man often gives his sister in exchange for the woman he has married. The debt incurred by marriage is thus a reciprocal matter between brothers-in-law, who form a natural pair of 'twins'. There is, however, a noticeable patriarchal trend among the Urubus, and it is perhaps for this reason that in their version of their twin myth Mikur mimi is killed off: in the Tupi version (based on a less age-conscious society) Mikur mimi was allowed to live, even being acknowledged by Maír as his true son.

There are thus living examples of Self and Other round which the figures of the mythological twins can take shape. In horde societies, however, which seem to be confined to the southern pampas region, relatives may not marry, and men must take their women from outside the horde. In myths of these societies the figure of Maír is present and indeed exaggerated, while Mikur and his son are absent, for there are no ambiguous relationships based on in-lawship within the horde: all the tension is between one horde and another. The relations between hordes are however not set ones, such as exist in tribes with moiety organisations, where men of one moiety can only marry the women of the other. Tribes with moieties cannot allow relationships to be ambiguous, they must be formal, and the moieties equal: and it is in such tribes that Maír and Mikur, sometimes known by other names, are on a more equal footing.

The myth of the twins can be brought even further down to earth. Maír mimi's victory over his father is an abstract and elemental one, and its only sign is the noise of the thunder: he has found no other treasure in his journey through the underworld. The hero who descends into the underworld and returns to the light of day bearing a treasure is a universal figure in myth, often symbolised by the sun. Among the Urubus, the hero's name is Turiwar, and the treasure he has found is the feather headdress—that is, the sun itself which he had become. The myth of Turiwar, to be told in the next chapter, begins to take us into the world of rite.

CHAPTER EIGHTEEN

A Cannibal Tale

ONE day Antonio-hu and the other men came back from hunting, having killed three deer. That night the stink of cooking meat spread through the village, and we ate till we could hold no more. There was so much meat that we couldn't finish it: word went to the neighbouring villages, and the next day a stream of visitors came to help us with it. They sat in our hammocks, with great chunks of meat in their hands, exchanging gossip at the tops of their voices, and made such a noise that Antonio-hu and I moved to a smaller hut at the edge of the village where we could talk in peace.

We were feeling rather lazy. Though I had eaten so much, Antonio-hu insisted that I should have a final delicacy, a deer's brain; so I sat on the ground, scooping it out of the skull with a sliver of wood. Antonio-hu sat on a palm spathe, scratching his foot and searching for a jigger in it: Chico sat with his back against a house-post, looking at a tame parrot that preened itself on the rafters.

"Tell me," I said, as I finished eating the deer's brain, "tell me about Turiwar."

The story came fitfully. Antonio-hu kept on looking for a jigger in his toe, and our assurances that there was nothing there only annoyed him. Saracaca came with a pot full of shibé, made from newly roasted manioc flour, which had to be sampled; and then one of the visitors came to ask me for tobacco. Antonio-hu only got warmed up towards the end, when he had some acting to do.

There was once a man called Turiwar, he said, who wanted to kill an aé. The aé is a mythical animal that is supposed to look like a very large spider monkey, with red hair, red eyes, a blue penis, and blue bones. It's a *tenetehar*—that is, sometimes it's just a monkey, and sometimes it's like a man: a were-animal. It has attendants in the form of kinkajous, small, furry arboreal animals of which the Indians are considerably frightened: they say that if you're sleeping in the jungle, and a kinkajou defecates on you, you'll die. Brazilians also fear this animal, though for different reasons: kinkajous, they say, are sodomites, and a man sleeping in the jungle should protect himself with a well-placed cork.

Turiwar wanted to kill an aé because of its blue bones, which he planned to cut into beads and string into a necklace for his wife. He climbed a yupu-i tree—a tree whose fruit are much favoured by macaws—and waited there. It was full moon; when the moon stood in the zenith, an aé came, carried by a wind that blew through the tree-tops. The aé settled in the branches of the yupu-i tree, picked the flowers and began to squeeze their juice into a gourd bowl it carried. 'Ah mandiakwer!' it said as it squeezed. (Mandiakwer is the name of a soup made from sweet manioc: the yupu-i flowers were the aé's sweet manioc.)

Turiwar drew his bow and shot at the aé. The arrow went straight in the middle of its chest, and killed it outright. Turiwar cut the body open, took out the bones and carried them home.

His sister-in-law was very jealous: she hadn't got any blue beads, so she pestered her husband to kill an aé for her. Turiwar told his brother what to do, how he had to wait for the full moon, and shoot for the middle of the aé's chest. But his brother was a bad shot: his first arrow caught the aé in the right shoulder, the second in the left. He had no time for a third, for the aé was on him with a club, broke his head open, and carried him to its home underground.

When the man didn't come back the next morning, his wife got anxious and consulted a shaman. The shaman took his pipe, smoked it, fell into a trance and saw his familiar spirit. The spirit told him exactly what had happened, and what to do about it: when the shaman woke up he got everyone in the village to help.

Aé (the story now demands that the aé be called by a proper name) lived underground, and the entrance to his village was by a small ant-hole, such as that made by the tô-ô ant, who only lives in windy places. The villagers found this ant-hole through a spot of blood left at the entrance. They began to dig there, and at last came down to another world, a jungle through which ran a broad path. While the other villagers lit a fire above, and boiled large pots full of water, Turiwar descended and walked along the path towards Aé's village.

There was no one there except Aé himself. All the others were out hunting; they were all jaguars, except for Hai-hai the wild dog. Aé greeted Turiwar politely. 'Have you come?'

'I've come.'

After Aé had told Turiwar what the others were doing, he suddenly remarked that he had killed a man just recently. Turiwar saw this was true: Aé was dressed in his feather regalia, sitting in his hammock with a small palm-leaf mat beneath his feet, and his legs were bleeding from long scratches. (A man who has killed another, and is undergoing a ritual seclusion, cannot stand on the ground lest he fall ill, and he scratches his legs to mark the end of his seclusion. The myth compressed things a little, but was still accurate.)

Turiwar was frightened: to show just how frightened, Antonio-hu screwed up his thumb and forefinger to demonstrate the size of Turiwar's anus, and then circled his little finger at the top joint, as a mark of how small Turiwar's penis was. And he became even more frightened when Aé asked if he'd like to see how the Aé and the jaguars danced when they had killed a man.

Antonio-hu could not resist doing the dance himself. He looked round for something to serve as a club.

"Take this," I said, pulling a stick out of the fire. "Mind the end, it's burning." He tucked it under his arm, stamped the ground once or twice reflectively, then straightened up—"Ah!" he said and hurried off to get a headdress, adding some tall red macaw tail-feathers at the back.

"There," he said, highly pleased, and put it on. He began to dance, the butt end of a tawari cigar in his left hand and the smouldering stick, held club-wise, in his right; his back bent, and his right foot rhythmically stamping the ground—thump thump! And he sang a horrible and mournful song, with dismal wavering notes in a deep bass, as he continued to stamp, thumping his right foot heavily in the dust, bringing the left foot up to it, hopping forwards another pace, and thumping again.

"Look at my feet!" he called out suddenly, speaking as much for Aé as for himself. He went on dancing, but now balanced on his heels and raised his toes, first those of the left foot and then those of the right, wiggling them up and down—a trivial gesture somehow full of the most awful significance. Then he passed to the next stage of the dance.

"Aé passed the club round his head," he called. He stood up straight, passing the club over and under his left arm, on each side of his neck—an awkward gesture—on either side of his waist, and between his legs. "The club goes here, and here," he called out, and then he was dancing again.

Everyone in the village by now knew something was up, and were watching Antonio-hu with as much pleasure as

I was. Wirakangupik's wife Mirijun suddenly shouted from the door of her hut—"Show him how Aé was clubbed to death, show him, show him!"

"All right?" asked Antonio-hu of his performance, grinning at me.

"All *right*," I said.

He then started on a ludicrous parody of the previous dance. Aé had taken off his feather ornaments, put them on Turiwar and handed over his club—'now,' he said, 'you dance!' But Turiwar was sick with fright, for he'd seen a large pot boiling away on the fire, in which his brother's head was cooking: so his voice quavered as he sang, and his legs would hardly obey him. It was at once painful and comic to see Antonio-hu imitating Turiwar's dance, as he shuffled through the dust, his buttocks sticking out and his knees bowed.

It was at this moment, when Turiwar was dancing, that Aé noticed the other men of the village returning from the hunt. Aé looked at Spotted Jaguar as he came, and saw he'd killed a peccary; at Black Jaguar, who'd killed a tapir, and Red Jaguar, who'd killed a deer. All the time Turiwar was dancing, making use of the distraction to get behind Aé. Once there he raised the club he was dancing with, and brought it down on Aé's head.

Antonio-hu let himself go. "TOOM," he cried, as he gave a great swipe in the air with his stick. "*Toom!*" He swiped again, and then struck at the palm spathe on the ground which he'd been using as a seat. "Toom, toom," he shouted; over-balanced, knocked his head against the roof, jumped backwards over the palm spathe, and struck again. Then ran, ran out of the hut, bent double, with his arms tightly held to his sides; giving prodigious leaps into the air at every five or six steps, and grunting as he went. "Hu! Hu! hu-hu!" He went out of sight behind one of the huts, rounded some manioc bushes, and reappeared still shouting and grunting, halloo'd on by everyone, visitors included: "ooeh, ooehh!"

Back he came, pounding up the path and leaping over fallen logs, till he tumbled back again into the hut. "Off ran Turiwar," he said, still intent on the story, "off he ran, with all the jaguars after him. Hah," he sighed, taking off the headdress and pulling his shirt over his head. "Hot!"

Turiwar ran down the path, and soon arrived at the hole leading to the upper world. His friends had let down a liana rope with a noose at the end; he put his hands in the noose, tugged, and was hauled up. The jaguars sprang after him one after the other; they were angry, for not only had he killed their tushau, their war-chief, but he had made off with the headdress, the bone-flute collar, and the other feather ornaments, the only ones in the village. They sprang up, clawed his legs and bit his buttocks, but each time they fell back because the men on top were pouring boiling water down on to them. Soon Turiwar was on the world's surface again, and all the jaguars were lying dead at the bottom of the hole. The shaman came next morning, and had the hole stuffed up with tree-trunks; then he smoked his cigar over it so that it vanished, and there was nothing to be seen but a patch of jungle floor, looking as it does everywhere: carpeted with leaves, and thick with undergrowth.

This myth can be understood without much trouble. The Aé seems to be based on the kinkajou, which is nocturnal and, though it can be tamed, is often surprisingly fierce, especially if it has been made drunk. The Aé also shares attributes with the macaw, since both he and it feed on the same fruit; with the difference that while the macaw is an emblem of the sun at noon, Aé is a figure of the night sun who lives in that supernatural world presided over by the full moon.

He rides on the wind, and is at bottom the same as

Iwitu ramuĩ, Grandfather Wind, who got so angry when Maír mimi tickled his wife. Aé is sometimes confused with Timakanã and Curupir, since all three are responsible for odd noises in the jungle, especially when the wind makes the trees creak and groan; and he is indeed an anti-Maír figure like them. But he is more dangerous than they, since he wears a feather headdress just like Maír: he is a tushau, or executioner and war-chief, with magic powers. He is thus not so much an anti-Maír as Maír seen in a bad light, the Maír who lives underground.

There is a perfectly good underground Maír in real life: the jaguar, whose favourite den is in a cave, or among rocks. Jaguars have already appeared in the twin myth, where they ate Maír's wife; there is, besides, another story that tells how a jaguar boy tried to overcome Tupan the thunder, like Maír mimi, but failed and was killed. There is always more than one solution to any mythological problem, which is why mythologies sometimes appear to be inconsistent: for the consistence they aim at is that of the recurring situation in which people find themselves, now on the victor's side, now on the loser's. The jaguar is how Maír appears to the unsuccessful hero, and how the unsuccessful hero appears to Maír.

The situation in which two hostile males confront one another is a common one, whose outcome the hymn describes:

> Conquering Kings their titles take
> From the foes they captive make.

This method of succession to the titles and therefore powers of the defeated implies a useful mythological axiom, namely that the kings and foes who fight together have at bottom the same nature: the successful jaguar will become Maír. They are opponents because they want the same thing, and share the same attributes. The difference is that the victor displays these attributes as positive qualities, the vanquished as negative ones: Maír is a man, jaguar-fierce and stone-hard, while his opponent is merely jaguar and stone.

Aé both is and is not Maír: he is Maír in that he is tushau and wears a headdress, is not Maír in that he lives underground. The underground world is really a female place, the place where masculinity is most naked: all the unforseen consequences of male pretensions appear there, everything that a man refuses to be. It is a world of death but, being female, is also a place of birth, and the thing which is born is symbolised by the treasure the hero brings out with him. In Urubu myth death overtakes Turiwar's brother, just as it overtook Maír mimi's; and the thing born is the headdress, that sun of feathers.

Aé, who is the guardian and inhabitant of the female underworld, is the night sun, Maír in reverse. He could well be Maír himself, but only if he was turned inside out and made to show the celestial blue of which his bones are made. I know of no animal with blue bones except the pocapoca frog; its bones are made into beads by the Tembés and strung round children's necks, to make them sleep well. But this hardly shows the significance of blue, a colour the Indians love above all others. It is as though Aé had two treasures: one, its blue bones, a kind of emotional riches Turiwar could significantly give to his wife, to make her happy; two, his feather ornaments, which a man keeps for himself. Women go to heaven too, but only men can be like the sun.

The really important feature of the Aé story, however, is that it begins to show how mythology is acted out in real life. The incident where Aé dances in front of Turiwar shows how the Urubu executioner dances in front of his victim, calling attention to his waggling toes in order to distract the victim and make it difficult for him to dodge the blow of the club: while the head of Turiwar's brother, cooking on the fire, shows that the cannibal world hinted at in other myths really exists.

The execution of the prisoner, and the eating of his flesh afterwards, are meaningful acts only in the context of the proper ceremony. The true man among the Indians is hard and strong, tolerating no effeminacy; he is an implacable warrior, obstinate in revenging the deaths of his parents and relatives; but above all he is one who delights in ceremonies. It is indeed in ceremonies that the hero most displays himself, since his attributes are then manifest: moreover, in such a world the man who ritually acts out heroic virtues ends by becoming a hero.

The acts of the hero are all mythologically described. The Indian projects this description outside himself, and sees everything in its terms; he thus performs literally acts which might well be taken metaphorically. All Indians try to model their lives on that of the hero, whether to be Maír mimi or Turiwar; but instead of making the descent into the underworld themselves, they send down a representative in their stead. This representative is an enemy prisoner who is really, if ritually, killed and eaten—the fate that nearly overtook Turiwar.

The idea of a ritual execution still excites the Indians a great deal, though they are wary of talking about it in front of strangers. Of cannibalism it is almost impossible to get them to talk: a direct question brings out merely a bored and aloof denial, which is a bit too bored to be convincing. Pari, when I asked him, as well as looking bored, would look disgusted, and I soon gave up the direct approach.

One of the books I had brought into the jungle was the account of a Frenchman, Hercules Florence, of a journey through the Brazilian jungle in the eighteen-twenties. He was an excellent draughtsman, and the book was full of pictures of Indians in all their feathers and paint. The book was a favourite wherever I went, partly because there were some drawings of completely naked women, which were always turned to first and blackened by fingers demonstrating the point of interest; but mainly because the pictures showed Indians who, though recognisably caá-por or true Indians—that is, wearing feather headdresses and lip ornaments—wore headdresses and carried bows and arrows quite unlike anything seen locally. Their way of painting themselves was strange, too: which so delighted Saracaca that she painted her arms and legs in the style shown, and kept it up all the time I was there. She looked very pretty, as though she was wearing gloves and stockings.

Antonio-hu was more interested in the weapons, and at every picture he would turn to me and ask what tribe the man came from, and where he lived. At last, as we were looking at some drawings of Apiaca Indians, he pointed at a lance which was figured there and told me its name—"murussu," he said.

"Murussu," I repeated casually. "They used to kill people with that, and then eat them."

"Eat people!" said Antonio-hu in disgust, "this book's no good, throw it away, put it in the fire!"

But he went on looking—now, as I turned the page, at the Munduruku Indians, who he miscalled Timiruku.

"They also ate people," I said, to stimulate him, though it was quite untrue. "They're angry, angry . . ."

"It was them that ate yandé ramuĩ," he interrupted, staring fascinatedly at the page. "Oh, but I'm sad, looking at the people who ate yandé ramuĩ."

"They did? Who else did?"

"Juru pihun," he bellowed, after thinking about it a little—the Black Mouth Indians. "*Juru pihun!*"

I turned the page back to show a picture of a Munduruku who had a great black stripe over his mouth, like a slice of water-melon, painted from ear to ear. "Look, Juru pihun!"

Antonio-hu looked silently for half a minute. "They ate yandé ramuĩ," he said in a low voice. Then in a shout, "Burn the book! it's no good!"

He remembered other tribes who were cannibals, the Wira tu, the Merikern. "Three," he said, tucking his forefinger underneath his thumb and holding his hand up with the other fingers straight—"Three! They ate yandé ramuĩ, knocked him on the head, *paam!*, and ate him."

I mistook what he said. "Yandé ramuĩ ate people?"

"No! No!" said Antonio-hu in a harsh voice, "white men are always telling us that we ate people, that we ate whites. It's a lie, we didn't. We didn't eat people. Never! Juru pihun ate people, but we're good, we didn't." After a little he added, "Capiwan ate people."

Capiwan, as I luckily knew, was the name of a sub-tribe of the Urubus; a name that referred also to the tushau, the war-chief, of that sub-tribe. Here at last, I thought, was the chance for me to find out about Urubu cannibalism without seeming to, without hurting Antonio-hu's feelings. So I jumped up excitedly and said that I had heard an account of how Capiwan killed and ate people, which I retold as best I could. The account was really that of Tupinamba cannibalism: I described how the Tupinamba used to go raiding, catch a prisoner, bring him back to the village, tie him up and then club him to death.

"Yes, yes," said Antonio-hu, shifting restlessly in his seat. "Capiwan used to kill our forefathers and eat them . . ."

"How?" I asked insistently, "how? How did Capiwan kill them?"

Antonio-hu suddenly came a to decision. Next to him, on the bench we sat on, was a young boy. Antonio-hu seized him by the arms and then shouted—"They tied their arms behind their backs! They tied them with a cord round the waist! Then . . . paam!" Antonio-hu struck with an imaginary club, then teetered on the seat with his head lolling to one side, like a dying man. "Sometimes they put a cord round their necks and *strangled* them!" he continued in a gross voice, clawing graphically at his throat.

Alishandre, who had come over from his side of the hut to see what was going on, leant right over and spoke into Antonio-hu's ear. "Show him how it was!" he said.

The rope that was tied round the prisoner's waist must have been some twenty or thirty feet long, and the other end was tied to a house-post. The prisoner could thus still move about fairly freely, and when the executioner approached with his club he would do everything he could to avoid being struck. This is what Antonio-hu now showed us: he got up brusquely, sweeping the bench clear of people as he did so, went outside the hut and then suddenly rushed back again, ducking under hammocks, whirling around and jumping from side to side as he imitated the prisoner ducking the executioner's club. I leapt up too, to hold the hammocks out of his way, and he continued his act: he side-stepped, ran to and fro, brought himself up with a jerk at the end of the imaginary rope, ducked his head rapidly from side to side, jumped, and twisted round; he rushed in and out of the hut several times more, spun round violently, ducked again and then suddenly stopped and laughed.

"Paam!" he gasped breathlessly, striking himself on the side of the head. "Dead."

"Tell him about Capiwan," Alishandre urged; "tell him how Capiwan killed yandé ramuĩ."

Antonio-hu started the story before even getting his breath back. "Capiwan, Capiwan," he said, panting, "once yandé ramuĩ was over there, visiting Capiwan . . ."

Once yandé ramuĩ had paid a visit to Capiwan. He

and his wife, yandé ari, our grandmother, were alone; his brothers had not liked Capiwan, they'd told him 'You're no good, you eat our brothers and our children. We're not staying with you any more, we're going home.'

Yandé ramuĩ stayed behind for four, five months. Then Capiwan said, 'Don't go away now. Stay with us, marry my daughter.' Capiwan gave him a hut as well, there yandé ramuĩ lived with his two wives.

One day Capiwan told him, 'Go and get mandiakab, sweet manioc.' Yandé ramuĩ went, yandé ari went, Capiwan's daughter went. When they were in the roça, yandé ramuĩ called out to Capiwan's daughter, 'Come and suruk!' he said. Capiwan's daughter didn't want to, she didn't come. 'Come and suruk!' She didn't come. 'Come and suruk!' (Antonio-hu rubbed a finger slyly in his palm.) Capiwan's daughter didn't say anything, she didn't come: she went away, back to the village. 'Don't go back!' shouted yandé ramuĩ, 'your father will be angry if you haven't brought him mandiakab!' Capiwan's daughter went back and saw her father, she told him that yandé ramuĩ three times had tried to shoot her with a takwara, and each time she'd jumped and he'd missed. Bad woman, she lied! Capiwan didn't say anything, but called to yandé ramuĩ. 'You're bad!' he said, 'you tried to kill my daughter.'

'It's a lie,' said yandé ramuĩ. 'I wanted to suruk her, I called three times, Come and suruk! and she wouldn't.'

'No,' said Capiwan, 'it wasn't your penis you tried to stick her with, it was a takwara!'

'Your daughter's no good, she won't let me suruk her. You gave her to me but she won't let me.'

Capiwan ordered his people to make cauin. He wanted to get yandé ramuĩ drunk, so that he could kill him. So they made cauin, three huge pots of it, and everyone drank and drank. Capiwan took a look at yandé ramuĩ and thought, Now he's drunk! 'Tie him up!' he said to his people.

'What are you doing, tying me up?'

'You tried to kill my daughter, now I'm going to kill you,' said Capiwan. And he tied him up firmly with his wrists behind his back, and a long cord around his waist with its other end attached to a house-post. Then he took the club and struck at yandé ramuĩ . . . wshhh!

"Hunh!" grunted Antonio-hu, twisting violently on the seat and ducking out of the way of an imaginary club. "He missed!" And Antonio-hu, in love with his demonstration, hissed, grunted and ducked for several minutes as he played executioner and victim.

They started when the sun was just above the horizon, every man took the club and struck at yandé ramuĩ, and they missed, everyone missed. When the sun was coming down the sky again, in the afternoon, yandé ramuĩ was tired, tired, he was hot and running with sweat. 'It's because I'm drunk that you can do this to me!' he shouted, 'if I wasn't drunk it would be me killing you!'

At last everyone in the village had tried to kill yandé ramuĩ, everyone but an old man. Capiwan said, 'Go and get the old man, bring him along to kill this one here!' They brought the old man along, holding him up by his arms: the old man took the club and lifted it, he lifted it over his head and struck. But he didn't strike. Yandé ramuĩ ducked, hunh! but the old man hadn't struck. But when yandé ramuĩ ducked, and put his head on one side, then the old man struck, wshhh . . . paam!

Antonio-hu put a hand just behind his ear. "Paam!" he exclaimed, "yandé ramuĩ was dead!"

Yandé ramuĩ fell dead to the ground. 'Ha ha ha ha ha!' laughed Capiwan, 'now he's dead, that one who tried to kill our sister. Cut his head off!' They strung yandé ramuĩ up by his feet, cut off his head and out came the blood, ti ti ti ti ti susssssss, blood! they caught the blood in a pot and Capiwan mixed it with manioc flour and ate it. Then they cut up yandé ramuĩ and cooked the meat, they ate it, all of it . . .

"How did they cook it?" I interrupted, "did they smoke it?"

"No," said Antonio-hu scornfully, "you boil it, you cut it all up and boil it, boil it, boil it, you don't smoke it. Boil it, that way it gets all tender."

At this moment Chico, who really had no dramatic sense, said he wanted to go and have a bathe. Admittedly it was seven o'clock in the evening, we hadn't had supper, and Antonio-hu had been talking since three in the afternoon: but still. "Is that all?" he said.

"Enough?" Antonio-hu asked, always courteous.

"Not enough," I said, "go on!"

Yandé ramuĩ's relations, Antonio-hu continued, used to come from time to time to see if he was all right: someone would climb a tree at the edge of the clearing and look across into the village. They came again, they climbed a tree and looked—they looked, where was yandé ramuĩ's hut? There was nothing there at all, and the village had been abandoned. The men went back to their village and told the tushau. 'Yandé ramuĩ's been eaten!' they said, 'Capiwan has eaten him! There's no one there at all now.'

'They'll come back,' said Tushau, 'in a month they'll come back to make another cauin.'

Next month Tushau sent all his men to Capiwan's village. He told them what to do. They went to the village, and someone climbed a tree and looked across. There was Capiwan, making a cauin. They circled the village and then they began to snort and grunt like peccary—wah wah wah wah, hrrn hrrn hrrn! Capiwan was there at the cauin. 'Peccary!' he said, and sent his men out to hunt them. Wah wah wah, hrrnhrrn! Tushau's men began to tap their bows with sticks, to imitate the peccary gnashing their teeth. 'Where are the peccary?' said Capiwan's men, as they ran into the jungle. On they came. 'Let them come close,' Tushau had told his people, 'till you can see them properly.' On they came: Tushau's men saw them properly.

"Pik!" said Antonio-hu, drawing an imaginary bow-string to his ear and letting it go. "Pik! pik! No peccary there at all! pik! arrows!"

Antonio-hu fought out a minor battle by the light of the fire, all by himself. "They fell dead," he cried, "the arrows pierced their backs, their shoulders, their legs! Pik! pik! The arrows came from all round, they all pointed into the middle of the village!"

There in the village the women heard the shouts: they ran out of their huts into the roça to hide between logs and manioc plants, but the arrows found them out, arrows in their backs; they crawled under fallen trees, pik! arrows in their arse-holes. Tushau's men killed everyone.

There in the village they looked for Capiwan. They were angry. They bound him with a rope, undid the string round his penis and then they twisted his penis round and round, they pulled at it, jerked it, they took their takwaras and stabbed it through and through. 'You killed our brother!' they said.

'No I didn't! He did, he killed him!' and Capiwan pointed up to the roof where the old man lay in his hammock. Paam! They killed Capiwan with one blow of a club, *paam*, he was dead. They climbed up to the old man in his hammock and tipped him out so that he fell on to the ground. 'You killed our brother!' they shouted at him, and they rolled him into the fire and covered him with embers, they burnt him all over and then they killed him with a club. They burnt the hut over the bodies, burnt it completely so that there wouldn't be anything left of the old man who'd killed yandé ramuin, nothing left of Capiwan.

Yandé ari was still there. She went back with the men, back to their village. She was sad, sad, she'd seen her husband being killed and eaten. Back in the village Tushau was waiting. 'I'm tushau!' he told her. 'I'm young, marry me!' She married him. 'It's all right for people to eat peccary,' said Tushau, 'peccary's all right, deer, tapir, paca, agouti, that's all right; jaguars, no, you don't eat jaguar's, you don't eat anteaters . . . men you don't eat! You don't eat men!'

"We never ate people," said Antonio-hu self-righteously, "only Capiwan ate people."

"What was the old man doing up there in the roof?" I asked.

"Capiwan told him, 'You mustn't walk about the village, people will be afraid. Don't walk about the village, don't go hunting, stay there, in your hammock there under the roof and we'll bring you food, shrimps, fish. Don't get up, don't walk about!' "

Chico took the opportunity to go and have his bathe and left, despite Antonio-hu's attempt to dissuade him. "It's dark, it's cold, wait till morning!" But Chico had had enough—not that it did him much good, for Antonio-hu was big with another story when he returned, which lasted till nine o'clock.

As for yandé ramuĩ—Piriangwa was his name—his death was a classic one. War and marriage are things which go closely together—indeed, the old Tupinamba Indians called both brothers-in-law and enemies by the same word, tobajara, which etymologically means to be face to face with, to be opposite.* With war and marriage go trade and barter. In parts of South America there are tribes, originally hostile, who have made peace with one another by becoming involved in barter, so closely sometimes that they barter that most precious of all commodi-

* The Urubu terms are similar: hẽ ruwaiyar, my brother-in-law; hẽ ruwaiyan, my enemy.

ties, women. Thus may start one of those tribes divided into moieties, the men of one moiety only being able to marry the women of the other. Such a tribe has brought the war inside its own boundaries, and during ceremonies the moieties confront each other and may even stage a mimic battle.

Among tribes without a moiety organisation, the feeling that barter is a kind of war (just as war is a way of getting what you want, especially women, without exchanging anything for it) is most felt between a man and his in-laws. A son-in-law has always some obligations, though he can get out of the worst of them by marrying into his own family. Such close marriages usually work very well: there is little chance of a man being browbeaten by his father-in-law, or a wife being bullied by her husband, if they are constrained or supported by other relationships. In-laws who are strangers, however, may not fare so well: if they quarrel, as Caro did with Toí (Chapter 9), the lack of any constraining relationship outside marriage may make a feud inevitable. What happened to Piriangwa is an example of this: he was a brother-in-law who was also a stranger, so that when he affronted his wife and his wife's parents he had no one to back him up. Capiwan took advantage of the situation to have a cannibal feast. This is quite natural: the Indians' world is one where mythology does not have to wait upon set occasions to be acted out, but is suddenly incarnated in what seems to be trivial and everyday events. Prisoners are normally executed only after long preparatory rites, as will be seen, which give them the mythological attributes required of them; Piriangwa could be killed and eaten without these rites since his daily life in the village, and his eventual provocative act (whatever it was), were the exact if unconscious equivalent of rite. Since the supernatural world provides the motive and reason for this one, it is not surprising that the story of Piriangwa, though a piece of history, can also be read as a myth.

CHAPTER NINETEEN

The Ritual Execution

One could always tell a chief in the old days, because he wore his headdress and his other feather ornaments every day, taking them off only when he went hunting. The other men wore theirs only on special occasions, at feasts or when they were going to make war. They needed them then in order to become what the ornaments showed they were: every man a Maír.

When they went to make war, the men put on their headdresses, tied on their armlets and wristlets, their belts and their collars, and painted themselves. They painted black stripes and circles on their faces with genipapo juice, and blackened their mouths with the burnt flesh of electric eels; they painted stripes and circles of genipapo on their chests and down their arms, daubed a large stripe of sticky milk from the massaranduba tree from armpit to waist, and powdered it with charcoal from the spoonwood tree; they stuck tufts of white falcon-down on their bodies and in their hair. Then they sang.

They sang u-apik kangwer—u-apik meaning he sits, or puts down, kangwer meaning bones free from their flesh—a song, Antonio-hu told me, that should always be sung with a machete carried under the arm. The men all put leaves on the ground and sat on them while singing about the fierce warrior sitting in their place: the fierce warrior who was, of course, themselves. They would sing till dusk, when it was time to move into the jungle and surround the enemy village for a dawn attack: and their last song would be ningar puku, the long song, which Antonio-hu demonstrated for me.

He was a bit put out when I asked him to, for he hadn't a club, and ningar puku should be danced with one. We found him a stick, which he tucked under his armpit, the butt held in his left hand, palm facing in. He murmured a few bars to himself to make sure of his voice, and then he was off. It was one of those unspeakable whooping songs which the Indians delight in, the words quite unintelligible—if indeed there were any words, for Antonio-hu could not tell me of any. It was just ningar puku, he said. But he danced, a dance of great verve and bravura, lifting his feet alternately high into the air and stamping them down with enormous decision. He stopped halfway through, and called for a headdress: with that on he began again, now with real panache. He changed the stamp to a kind of goose-step, swinging his legs out and kicking the dust with his toes, then stamping again; all the time with his back somewhat bent, and singing barbarously loud.

"Ningar puku!" he said emphatically, straightening himself up. "You sing, then you run into the woods to find the Guajajas, your right hand on your head, your finger up your arse-hole. That's what yandé ramuĩ told us, he said—'If you don't put your finger in your arse-hole you won't kill anyone.'"

Antonio-hu started to hum again. He picked up an empty tin and beat it gently while he danced a little; then he looked round and stopped.

"I'd better not sing any more," he said, "because of the women, their children won't be born." It was the camayi tiapiw maé song, the noise-thing song of Camayi, the cicada. If women should hear it, their wombs become so knotted that once pregnant they can never give birth to their babies; and even young men are affected, for their bowels begin to issue from their anuses.

"Young men can't listen at all?" I asked.

"They listen a little. They go off to the end of the

village, they listen a tiny little bit. The others, the men, they're all together, singing loud."

Chico yelled at the women to make them stop their ears—they none of them took much notice, except for Ari, Antonio-hu's daughter, who obeyed giggling—and Antonio-hu started again, this time with a great bundle of arrows under one arm, and the tin resting on top of them. The tin should really have been the shell of a yenda'i turtle, hanging by a strap from the shoulder, which makes a surprisingly noisy drum. The camayi tiapiw maé dance was somewhat ludicrous: Antonio-hu danced bow-legged, kicking the dust up in front of him with his heels, beating first the tin and then, because they made a better noise, the arrows, and all the time singing a mournful, high-pitched song.

One of the most difficult things to do is to learn about a ceremony which is no longer held. Stories, like the one Antonio-hu told me about Capiwan, hold a lot of information, but by no means everything; direct questions are not much use because answers to them leave out the context and the trivial detail which still may be important, and telescope events hopelessly together. I never discovered when camayi tiapiw maé was sung, or why; nor did I find out very much more about ritual executions and cannibalism. The little I did find out, however, is interesting, and becomes even more interesting when it is compared with the accounts of Tupinamba cannibalism, which are full of detail. (The best summary of them is to be found in Alfred Metraux' well-known *La Religion des Tupinambas.* The accounts themselves are nearly all first-hand reports by missionaries and adventurers, written soon after Brazil was discovered.)

The camayi tiapiw maé song is obviously important, since women are not allowed to hear it, and young men allowed to hear only a little. It is one of the few indications that Urubu men had to pass through an initiation before they could marry. There was no cult, properly speaking, with mysteries and secret objects (like those of the Jurupari cult among tribes to the north) which it was death for the women to see; instead, a man publicly executed a prisoner, and afterwards publicly performed the rites which gave him the strength of his own manhood. There were some secrets involved, but these—except for the camayi tiapiw maé song, and things like it—were secrets not of fact but of act. The fact of a ritual execution was plain enough; it is only when something is not quite real that it must be made secret.

The form which ritual secrecy takes among the Urubus is that of seclusion. It is a time when they are hiding things not only from other men, but from ghosts, spirits, and their own womanish impulses. But what an Indian hides is not something that he has made, but the making of that thing—especially if the making is difficult, and its outcome not sure. That is why Indians, when they want to make pots, go into the jungle by themselves where no one else can see what they are doing, and make them there, abstaining from eating, drinking, urinating, and sleeping with women. They make good pots, but large numbers crack in the firing: a fault they see as rooted in the spiritual act of making, not so much in the facts of how the pot is made, or what it is made of.

In the same way Indians hide from the world at important moments in their lives, by secluding themselves, eating a restricted diet, abstaining from sex, that most powerful of contaminants, and undergoing endurance tests. They have to do this because they are making themselves. A girl at first menstruation secludes herself since she is about to become a woman; a married couple, at the birth of their child, because they are becoming parents; a man, when he has ritually executed a prisoner, because he is transforming himself into Maír. Such a ritual seclusion imitates a stay in the underworld.

There are some things which can never be made public, even after the secret act of their making has been accomplished, since they are themselves creative acts. Hunting charms are such things: an Indian prepares them deep in the jungle by himself, wraps each in a leaf packet, and hides the packet in the thatch of his hut. If another man unwraps the packet and touches the charm, perhaps if he even looks at it, the magic is lost; if it is a woman who meddles, the magic may turn on her and kill her. These magic objects, which lose their virtue if they are not kept secret, are to a man what the ogre's heart in fairy tales is to the ogre: the heart, hidden in some unlikely spot, is all that gives life to the ogre's enormous bulk, and when it is destroyed, the ogre instantly dies.

The feather headdress is a piece of magic which works both in private and in public. Women may not wear it, or even tie it on to their husbands' heads, lest they have their eyes pecked out by yapu birds: it is therefore something of a secret from them. But it is always effective, no matter who sees or handles it, since its purpose is to display the force it emblemises: it is a piece of social magic, or ceremonial. But it also had to be made: and the story of its making is the myth of Turiwar and the Aé.

It is usually Maír who is thought to be the owner and therefore the 'maker' of the headdress, for Maír is the acknowledged culture-hero, and bringer of gifts. It is a paradox that these gifts are usually fatal to him who brings them. It is true that they are that which makes life possible for an Indian—his food, his language, his customs and ornaments—but their peculiar value is really seen during rites, when the gifts are used both to celebrate the occasion and, in a way, to provoke it. Culture-heroes always die at the hands of those they have made men, and no one is a real man till he has killed an enemy and fathered a son. The culture-hero's gifts are directed towards these two ends; the moment a man accepts the gifts he becomes responsible and can no longer avoid his fate, which is to be a man. There is no such thing as a free gift: to receive a gift always puts one in debt. The consequences of giving often become absurd, as in the potlatch ceremonies of north-west America, where a man who is given something will waste all his property in the attempt to give back more, and so humiliate the first giver.

The main effect of the culture-hero's gifts among the Urubu is to obsess them with virtues of hardness. All their masks, their paint, their feathers, their formal speech or ne'eng hantan, their ceremonies, are attempts to embody the hero's main gift of manhood; attempts which, being exaggerated, always tend to make the Indian overreach himself, to be harder than he need. This naturally breeds resentment in those who are over-reached, who immediately decide on the usual motive for Indian wars, and even for the ritual execution of prisoners—revenge.

Revenge is the motive for a ritual execution, but does not explain the rites involved. These rites are really part of the culture-hero himself, and his justification. He need not be a man: often he is a totemic ancestor—a bison, an ibex, a parrot, a witchetty grub—whose life and death are ritually acted out. Man is that paradoxical animal who does not know what he is, and who learns through the image of the ancestor in that most serious of all play, rite. Before this play, the ancestor did not properly exist: he was but one of innumerable images, none of which had a home. Since the ancestor or hero did not exist, Man did not exist either: without a firm image of himself to look at he is whatever impulse takes him, living in a state where not even demons have appeared. When, however, this image emerges, at one stroke it invents fathers and sons: it is the logic of paternity. Theologists, said Meister Eckhart, 'agree that God the Father conceived his own nature in originating the eternal Word and all creatures'.

The myth of the hero shows how fathers beget sons, and how sons become fathers in their turn. It is called a myth because it is a drama of images, a spiritual adventure making plain the shape and function of the mind itself: though to the Indians, incapable of abstracting the meaning from the physical act which displays it, the myth is the actual reality to be taken, in the words of St. Francis, 'literally, literally, literally, without gloss, without gloss, without gloss'. Myth thus literally acted out becomes rite.

It is however not enough to celebrate the hero whenever he is made to appear in rites: he must be present also in daily life, and a passage-way made for him from one celebration of his history to the next. This passage-way is marked in certain important places by means of negative signs or taboos. He who breaks these taboos finds that his image of the hero is, so to speak, lost in the other world and unable to find its way back, for the taboo is as good a barrier whichever side you come from. Soul-loss is a common fatality among primitives—Timakanã, it will be remembered, ate a man's soul—and its threat is present in mythology whenever the hero is transforming himself; if a man who is unworthy attempts the act, or sometimes if he merely witnesses it, he may be petrified with fright and turned into a stone, or he may give way to the impulses which the hero would have controlled, and be transformed into one of the animals who are emblematic of the hero's ruthless power—a jaguar, a cayman, a great anteater, a piranha.

Taboos impose a meaningful direction on him who obeys them, by giving shape to the dangerous impulses they circumscribe. If they are obeyed all the time, however, they defeat the hero's purpose which is to conquer danger, not merely to avoid it. Now the biggest taboo is on killing the hero, for to kill him casually means that you do not recognise him for what he is, while to kill him knowingly lays you open to all the forces the hero controls. The hero, however, must be killed, not only as a proof that he is alive, but in order that the man who kills him may gain possession of his attributes and thus become a hero himself. The only way of doing this is in a rite, and is the opposite of how you kill an ogre. The ogre is a body without a heart; the hero is a heart without a body. To kill him a body must first be found and the heart ritually invoked to come and inhabit it.

It is here that the accounts of Tupinamba cannibalism come in useful: through them one may see just how a body was found for the hero, and the hero then killed. 'Warriors who are truly brave,' the Tupinamba boasted, 'die in the land of their enemies': they harden themselves to face the worst dangers they know of, rather like Turiwar in the underworld confronting Aé. Indeed, one of the Tupinamba myths has Maír himself as a prisoner about to be ritually executed. It is the enemy prisoner who provides the body for the hero, and who eventually turns into a hero: his ritual execution is the underworld brought into the light of day, the occasion where he proves himself.

The Tupinamba used to go on raids to capture prisoners. If they were successful, and had killed some of the enemy in the battle, they would remain in the jungle to cut up the corpses, cook them over a slow fire and eat them there and then without ceremony; bringing back perhaps a basket or two of cooked flesh to be eaten by those who had stayed behind in the villages. The prisoners, however, they took back with them alive, killing them only at the end of long preparatory rites. Why, one asks, did they bother?

They bothered because the jungle isn't a place. The jungle is the land of animals and ghosts: the only places which are truly somewhere are the villages with their Maír-like chiefs. Thus Indians who eat human flesh outside the village are not really doing anything noteworthy, since the jungle is anyhow a place where people get eaten, one way or another—if not by men, then by jaguars or ghosts. Eat man's flesh in a village, however, and the action at once becomes enormously significant.

Once in the village, the prisoner is truly in the land of his enemies. Warriors crowd about him biting his flesh in order to mark what pieces they are reserving for themselves at the coming feast; the prisoner, as he approached the village, has already been made to cry out, 'I, your food, have come!' Ululating women rush out and dance about him, flutes made from the bones of other enemies are played in triumph, and the prisoner is led in front of the chief and in front of the sacred maracas by whose means the shamans have foretold his capture. His hair is then cut in the style of his captors, and he is painted as they are; then he is led to the grave of the man whose death will be avenged by his own, and made to clean or 'renew' it.

Here are more reasons for taking the prisoner back to the village. Just as the jungle is not a place, so an enemy is not a human being. The Indian knows as human only the members of his own tribe, and the eponym of the tribe is usually one just meaning 'Man'. To eat an enemy in the jungle, therefore, is hardly to eat a man at all. To do this he must be brought into the village, made to look like a man, and adopted into a family.

The 'renewing' of the grave is as much as saying that the man who lies there is newly dead, and that the prisoner is in some way responsible. He later takes full responsibility for the dead man, by being given his widow and all the possessions which were not buried with him. (If the dead man left no widow, the captor will give his own sister or daughter to the prisoner, and if the captor has neither, other men will think it an honour to give one of their womenfolk.)

The prisoner has now been adopted, and is free to walk about the village and even to go out hunting. He refuses, as a rule, even to try to escape, for if he did and returned to his own village his relations would abuse him for being a coward—'would we not have avenged your death?' they would ask indignantly. Some prisoners, however, did escape, often helped by the women they had been given as wives. If they were caught, both would be killed outright, the woman left lying in the jungle with her skull broken, the prisoner cooked and eaten.

On the whole, the prisoner's life was no different from that of the other Indians. From time to time, however, he would be dressed up in feather regalia, painted red and black and shown to the assembled villagers who reviled and jeered at him. The next day he would again be treated as one of the family. (Some prisoners, caught when very young, spent as much as twenty years in this kind of life before they were killed.)

The prisoner was an extreme form of *tobajara*, that word meaning both enemy and brother in-law. Coming from an enemy tribe, he was forced into a marriage and made a tobajara in both senses of the word. But he could also make love to any woman as long as she was not married. Certain facts suggest that not only the girls but the girls' parents welcomed his advances: for as a culture-hero-to-be he had fructifying powers which could be made use of. Indeed, making love is a pleasure that the hero can hardly be denied, and it becomes an honour to satisfy him. But one may not satisfy him too much: if he should make love to a married woman he is killed immediately, since a husband is also a Maír and a hero. The prisoner, however much he comes to incarnate Maír, remains a prisoner and an enemy, and any children he fathers are killed at an early age.

Thus the prisoner, though he has been adopted, is still an enemy: an enemy in the centre of the family. But he is only treated as one (except for those occasions when he is ritually jeered at) if he tries to escape: he then forfeits his adoption and is outlawed. A ritual escape indeed figures as a preliminary to the execution. The prisoner is taken into the open and encouraged to run away—'hide yourself! disappear!' they shout at him—so that he may be chased and recaptured as a man to whom no responsibility is owed. This ends his free life in the village: henceforth a heavy cotton rope is put about his ankles, and his owner stops feeding him. The effect of this is to make the prisoner go round the village stealing what he can and breaking what he likes. No one stops him. He is being made into a scapegoat, and the more damage he does the more justified will any revenge be.

It is the fate of the prisoner to act out a number of contradictory rôles and incarnate them in himself. He is an enemy who is adopted; he takes the place of the man in whose honour he will be killed; he is an in-law and an outcast; he is honoured and reviled, a scapegoat and a hero; he is intimidated but, if he shows fear, is thought unworthy of the death that awaits him. By acting out these primarily social rôles, he becomes a complete human being, exemplifying the contradictions that society creates: an impossible situation, which can only end in his death. The impossibility is exaggerated when he is charged, by ritual, with the powers and attributes of the culture-hero: he becomes the representative of the other world living in the centre of this one, a Janus figure too sacred to live with.

One of the rites that turns prisoner into culture-hero links him with the club by which he is to be killed, and in part with any dead chief. All three (and only these three) are daubed with honey and covered with green egg-shells by a company of women. Afterwards the women spend an entire night in their hammocks, singing lugubrious songs to the prisoner who is in their midst; and another night doing the same thing to the club—'singing it to sleep,' as they explained it. (At a chief's funeral, of course, singing is out of place: the women shriek and wail, to frighten away the ghost.)

It is a commonplace that the sacrifice and the sacrificer are spiritually identical with the god to whom the sacrifice is offered; one has only to add that this identity is shared by the weapon which effects the sacrifice. The club is a kind of spare body like the Australian's churinga and, of course, like the prisoner himself; it also needs to be charged with spiritual power, in order to be effective in the rite. The Tupinamba did it like this:

'There cometh an honourable judge or stickler of the new Knight that shall bee, and taking the Sword (the club) passeth it many times betweene his legges, putting it now on the one side, now on the other, even in the same manner that the Piper's dogs do pisse betweene their legges, and after taking it by the midst with both hands he pointeth it with a thrust to the eies of the man which is to die.' (*Purchas his Pilgrims*, book 7, page 1296.) This is just what Antonio-hu did, when he demonstrated the dance of Aé. The club is being given life by contact with a man's flesh. One may easily guess the symbolic nature of the club, after its entertainment by the women—the prisoner has been entertained in much the same way, and for a longer time. But just as the prisoner remains an enemy, for all he is a culture-hero, so the club remains the executioner's weapon for all its hardness and its having been put to sleep by women. It is the most abstract personification of Maír and his power, the only way in which the executioner, Maír-to-be, can kill the prisoner, Maír-that-is.

The prisoner's last day dawns with everyone singing round the club in the middle of the village. The prisoner is woken, the temporary hut in which he lived is pulled down and he led to the execution ground. There an enormous cord of plaited cotton, tied in complicated knots, is put round his waist and the loose ends given to his wife to carry. He has his last chance of avenging himself, which he does with insults and by throwing genipapo fruit and clods of earth at his jeering audience. In this he is helped by his wife. Then seven or eight old hags, painted red and black and wearing collars of human teeth, come on singing, beating out the rhythm on newly painted pots which are to hold the blood and guts of the dead prisoner. A fire is lit a couple of paces away from the prisoner, apparently as a reminder that the occasion is in honour of the fire- and sun-hero; then another hag enters running, bearing the club with its handle pointing upwards. This she gives to a waiting warrior.

Now the executioner arrives, with a red headdress and a cloak made from red ibis feathers; with shell and feather collars crossed over his chest, with bracelets and anklets, and a huge roundel of white rhea feathers over his buttocks. He dances round the clearing, rolling his eyes wildly and imitating the act of a falcon striking down its prey; he takes the club and then gives it to that 'honourable judge or stickler', either the war-chief or a man renowned for his bravery, who by contact gives it the virtues of his own body. 'Are you not of such and such a tribe, our enemies?' the executioner then asks, 'have you not yourself killed and eaten many of our relations and our friends?'

'Yes, I'm a great warrior, and truly I've killed and eaten many of you.'

'Now you are in our power, soon you will have died by my hand, soon you will be roasted and eaten by us all.'

'Very well. My relations will avenge me.'

The plaited cord is so tied about the prisoner's waist that both ends are free. Each of the ends are now held by two or three men: it is as though a tug-of-war is taking place, with the prisoner in the middle. (The Urubu simplify matters by tying the prisoner by the cord to a house-post.) The prisoner can still jump and duck from side to side, and the executioner is often hard put to it to kill him; sometimes, even, the prisoner will be given a club to defend himself with, or will jump on the executioner and wrest his club from him. But at last the prisoner is struck down, falls to the ground and has his brains dashed out. At once the women rush to gather up the brains, to dabble their fingers in the hot blood and lick it; encouraging their children to do likewise, and telling them that now they are revenged on their enemies. Nursing mothers would even smear their nipples with the blood so that their infants could share in the triumph. Then the women scald the corpse with boiling water and scrape away the outer skin. They put a plug in the anus so that nothing should be lost, and transfer the guts to a pot—'I leave you to imagine what a broth,' commented Thevet, the sixteenth-century chronicler. The body is jointed and laid on a moquem over a slow fire: as it cooks the women lick up the fat that dribbles on to the sticks, such is their appetite for human flesh.

The executioner goes immediately to his hut where he takes off his ornaments and passes through a strung bow that a friend bends for him, without touching either the bow or the string: as though he were passing through ita kitik. Once inside the hut he performs other rites to confuse and frighten away the prisoner's ghost. Then the prisoner's head is brought to him, and the threat the 'honourable judge' made with the club is carried out: an eye is gouged from its socket, and tied by its nerves to the executioner's wrist. This is the only contact he has with the dead man: he may eat none of the flesh.

This makes the meaning of the rite for the executioner plain enough. The sun, that emblem of Maír, is a symbol of consciousness; the eye is what sees it. What the executioner has done is to obtain from the prisoner that particular kind of consciousness which is the essence of being a hero, and which is the climax of the sun's passage underground.

The ritual execution among the Urubu must have followed much the same course. It was unfortunate that I learnt nothing more about the actual execution than I had been told in the story of Piriangwa and Capiwan; luckily, however, I later managed to hear further snippets of information about some of the rites that led up to and followed the execution. Some I got from Antonio-hu; but the most interesting I got from an extremely friendly and quick-witted Indian called Kuashi-puru, or Jumping Squirrel, who lived near the Tembé village of Jararaca.

The first thing Kuashi-puru told me was about the club: for I had asked whether the Urubus sang to it, as the Tupinamba used to. It turned out that they did, though in a quite different way.

"Yes!" he said. "A man brings the club from the jungle, where it's made. No one can see it being made! Everyone is in the house, drinking cauin. The man with the club comes—'here's the club coming!' they sing, 'here it comes.' There it comes among them. 'It's come, the club's come!' they sing. 'Your thing,' they say to the tushau. 'My thing,' he says. Then they sing—'his hand-thing waves about!' they sing." Kuashi-puru opened the fingers of his right hand, turned the palm this way and that, and waved it slowly, gravely, in front of his face.

" 'His hand-thing waves about!' Another man takes the club from the tushau. 'Your thing,' he says, and then gives it back—'my thing,' says Tushau. Everyone takes the club then gives it back to Tushau. 'His hand-thing waves about over the bowels,' they sing." Kuashi held an imaginary club over his right shoulder, then made a swooping motion downwards, up again over his chest, and followed the motion through till the club was over his left shoulder and his elbow under his chin.

"Everyone takes the club—'your thing,' they sing, and Tushau says, 'My thing.' Then he takes the club and puts it in the rafters, high up. There it stays, till it's needed."

The club is obviously an extremely potent object, for it is usually referred to as a 'hand-thing', and not as a club: rather as an Indian must not be addressed by his real name when you speak to him, but called 'fellow'. The point of this, presumably, is that the club is being made into something more than a mere club, so that to call it by name would stop the process of transformation: it would remain the club it was.

This is a consecration that takes place before the execution.* The rites that come afterwards are also consecratory, but they affect the participants of the rite, not the objects by which it is celebrated. The rite is felt to be so dangerous and important that the chief actor, the tushau or executioner, has to go into seclusion for many months: he stays in his hammock, which is slung high up beneath the rafters, eating fish and small shrimps, and may not show himself in public lest people fall ill, such is the spiritual power that invests him; nor may he tread on the ground unless a mat is put beneath his feet. Then a cauin is made, and on the day it is ready the executioner goes down to the stream to wash himself; then he returns to the village and stands outside the feast house, that large hut built especially for cauins. He faces east: as the sun rises some older man squats down in front of him and with a sharp squirrel tooth makes long cuts into his arms and legs, so that the 'bad blood' can come out. The

* The use of the present tense in what follows must be understood to refer to the historic rather than the actual present, for the Urubus no longer practise cannibalism.

women come with natural cotton and a decoction of cotton leaves steeped in water, to wash the blood away: for if anyone should tread on it he would turn yellow, like Mikur the opossum, and stay yellow for the rest of his life. After that all who have seen what has happened —and this means everyone in the village, for they are all watching—have their legs scratched in the same way. (This ritual scoring of the legs is also used to cure a malady called heavy leg, in which the limbs lose all feeling and become as heavy and insensitive as stone: the scoring, by letting out the blood, transforms this stoniness into a positive virtue. Turiwar had his legs scratched for him, after killing the Aé, by the jaguars who pursued him.)

Then comes the real test of hardness: like a girl undergoing her puberty rites, the executioner is made to wear two cotton bands, in which a mass of tapi-ai-ĩ ants have been inserted. One of these bands he wears round his forehead, the other round his waist. Wearing them he starts to dance, and a quantity of enormous wasps are thrown over him, adding their stings to the venomous bites of the ants. The man dances all morning, with the others dancing about him, till the wasps have exhausted themselves with stinging him, and the ants have begun to wriggle free from the meshes of the cotton bands; and all this time he must show himself to be hard and strong by not crying out with pain, and not trying to rid himself of the wasps and ants.

'I've just killed an enemy!' he then shouts, roughly seizing hold of some friend.

'You've just killed an ememy!' the friend shouts back. 'Give me your arrows.'

They exchange arrows, as men do when they are made chiefs, and continue dancing till nightfall when the cotton bands are taken off. The executioner is then given a large bowl of cauin to drink, and some stewed yashi tortoise to eat, to break his fast. Then the singing begins. The men sing how cauin is made: they sing manioc pounded in the trough, manioc put through the sieve, manioc roasted on the oven, manioc turned into cauin: they sing cauin muru, the cauin of praise. Then the women sing: they sing the food the executioner has had to eat during his long seclusion, they sing mushi-mushi ramuĩ, Grandfather Shrimp—how the roots of the pashi-i tree are rotting under water, and being eaten by the shrimps—they sing fish; and for some reason they sing mahau, the white deer that lives in the jungle, starting at every noise, looking from side to side. As they sing they dance: the men stamping, the women shuffling between the line of hammocks where the men are then taking their rest, and looking at their newly-scratched legs.

I am not sure if this sequence of events is correct, nor when the crowning moment of the ceremony takes place, but this does not matter. This crowning moment, as far as I can see, occurs the next morning at sunrise: Kuashi-puru again was the man who told me of it. I had made things easy for him by recounting everything Antonio-hu had told me about the ceremony, for at first Kuashi-puru had also been rather ill at ease when talking about cannibalism. Suddenly, however, in the most casual way possible, he told me that the executioner stood in the middle of the feast house, with everyone looking on, and undid the piece of string that tied up his prepuce. It took me a moment to understand the gravity of this act, for a man usually dies of shame if his prepuce is untied, and the glans penis is seen.

"Everyone sees? the women see it?" I asked.

"The women see. Tushau stands in the middle of the hut, *he* doesn't look at his penis, he looks away . . ." Kuashi-puru put a hand under his nose, and gazed abstractedly at the roof above him. "He doesn't look. The women all look, they come up and take hold of it."

"Ta!" said Chico in amazement.

"Yes," Kuashi-puru said gravely. "The women all come and hold the penis, they rub it in their hands; Tushau pulls back the prepuce for them, they rub its head. 'You're tushau,' they say, 'you're strong.' He doesn't look at them, none of the men look, they all look somewhere else.' Again Kuashi-puru put a hand under his nose, and stared at the roof.

"But, Kuashi," I said, "aren't the women ashamed to look? And Tushau, isn't his penis hard?"

"No, it's soft. The women want it, they like it—'Tushau,' they say, 'I want it!' 'No,' says Tushau, 'I don't want you to have it.' Tushau can only suruk his wife once a month, or he falls ill."

"And why do the women hold it?"

"It's a medicine," said Kuashi-puru, laughing, "medicine for headaches," and rubbed his head with his fingers.

"Kuashi!"

"Ya té! It's true!"

True or not, the actual form of the ceremony is extremely interesting. It is as though the executioner, after his long seclusion, is given back the penis which he has been unable to use for so long, in a public yet intimate fashion, while the women acknowledge him to be their man—*the* man, indeed, who alone has power over women and over the female underworld from which he has just heroically emerged. The women, as ever, are shameless in their recognition of this triumph and it is only the men who are ritually embarrassed, since it is their magic which is being publicly tested.

And what of the cannibal feast itself? The Urubu women were probably just as voracious for human flesh as their Tupinamba counterparts; the men and the children, however, must have had their share too. In this way they ate the prisoner's virtues and, by getting outside him, proved that they were larger than he was and so revenged themselves upon him. Here, of course, is another act of over-reaching, for which a vengeance would be exacted by the prisoner's relatives: for to be eaten is almost as bad as that humiliation Sir Thomas Browne wrote of—'To have our sculs made drinking bowls, and our bones turned into pipes, to delight and sport our enemies, are Tragicall abominations, escaped in burning burials.' And yet, being eaten has certain advantages. There is an interesting parallel between the custom the Tupinamba had of eating their enemies and burying their dead relatives, with that of a neighbouring tribe who ate their dead relatives but not their captive enemies. 'They hold it to be the greatest sign of love that they in this life can show to those who have absented themselves for the next, to give them sepulture in their bellies. The Tapuya in particular eat their children, if they happen to die shortly after being born.'*

They ate their children so that they might be born again. There is a general rule that he who eats man's flesh must beget that man as his child, one way or another: indeed, among the Tupi-Guarani the prisoner was reincarnated as the child of the executioner—that is, the executioner's child was given his name. Among the Tupinamba and the Urubu, things are not so clear-cut. The feast is called after the name of the prisoner, and a secondary feast, whose purpose is not quite clear, is held later in his honour; and though his flesh has been eaten by a hundred mouths, these mouths have but one belly between them—the belly of society, which can now use the heroic powers it has invested in the prisoner, and turn them to its own uses. The ogre has been killed without damaging his heart, which can now be used to re-invigorate society. Among the Tupinamba the 'heart' had a considerable effect. The executioner, if he was a bachelor, could marry; the widow of the dead man in whose honour the prisoner was killed could remarry if she wanted to; the executioner and his kin took new names; an old cycle of

* Vasconcellos, *Vida do Padre Anchieta*, Chapter 3.

revenge was closed, and the dead could rest in peace, while a new cycle (to be started by the prisoner's relatives) was about to open; and various political ends and friendships had been gained, by inviting neighbouring villages to the anthropophagous feast, thus making them as guilty of the prisoner's death as the village to which the executioner belonged. Besides this, the ceremony seems to have had some connection with the fertility of the fields. As for the prisoner, he had experienced his own transformation into a living and dying Maír, and so became one of the many heroes in the afterworld.

CHAPTER TWENTY

Leaving

"I'LL be sad when you go," Antonio-hu told me one day, "for then there won't be anyone to listen to my stories."

He had indeed told me a great number of stories, and enjoyed himself vastly in the process; and so had I, though it was hard work to remember them and write them down the next morning. There were stories of raids—one a considerable epic, for he took the best part of three whole afternoons to recount it—stories of ghosts, stories about Maír and the origins of things, stories full of scandals and gossip. It was a simple matter to start him talking: I already knew quite a lot about Urubu history and myth, and through my work on kinship had become familiar with the great names of the past, back to the fifth and sixth generations; all I had to do was to say, "Saé, fellow, tell me about the moon", or "Tell me about Kuashi-hu" —a great hero long dead—and out would come the story of the moon and his sister, or the fabulous exploits of Kuashi-hu who died fighting a neighbouring tribe. It was from listening to such stories that I learnt many of the details concerning cannibalism more accurately than if I had asked him straight out: for his memory, like that of all the Indians, worked best when dealing with particular occasions, and not with the abstract.

After about a month, however, I began to feel a certain indigestion of the mind, and began to have difficulty in thinking of subjects for Antonio-hu to talk about; and he started feeling drowsy, for during this time he had almost given up hunting, as a rule only going out with me to shoot monkeys with my ·22. He began to complain of pains in his arms and legs, and headaches, and general malaise, and he urged me to take out my hypodermic and give him injections in his arms, his thighs, his temples, his buttocks, his belly, everywhere. I gave him some aspirins instead, and told him to go and kill a tapir for us, but he said he was too weak to hunt. He became a little short-tempered as well, and as I seemed to have heard most of his stories, I decided I might as well leave.

The thought of leaving made me feel as sad as Antonio-hu, for we had become great friends, having all our meals together, a bond to which he attached much importance. One morning, when I was hard at work on my notes, Chico dished lunch up earlier than usual; and Antonio-hu was very hurt that I didn't immediately leave my work to come and eat with him. "If you don't come and eat with me," he said, "I won't ever come and eat meat out of your tins with you." I apologised, and hurriedly joined him, but later the same day I again offended him. His engaging daughter Ari—about thirteen or fourteen, with firm young breasts and a mischievous grin—brought us our usual afternoon shibé in a pot, from which Antonio-hu took his share in a gourd bowl. He raised it to his lips and then suddenly said, "I'm *sad*," putting the bowl down on his knees, the shibé untasted. "When you're gone, there won't be anyone to come with me to the stream, to bathe. When someone brings me a shibé, I'll be too sad to drink it, I'll be thinking how we drank shibé together. There won't be anyone to talk to, no one to sing with . . ." He dipped his finger in the shibé, stirred gently, raised the bowl again to his mouth and gulped. "Tobacco," he then said, simply, putting a hand out for it.

But my tobacco was nearly gone, and I told him so. So he sat brooding for half an hour, then rose, put a small wicker-work satchel over one shoulder and told me he was going off to the Parua river, a week's march away, to have a look at the women there. He gazed at me without expression, standing awkwardly on one foot. "Saé," I said, "if you go off to the Parua, I'll never see you again." At that he grinned sheepishly, took his satchel off and groped in my shirt pocket for my tobacco pouch.

We stretched out in our hammocks, and Antonio-hu asked Chico exhaustively who his relations were, going through the list one by one till Chico mentioned the name of his niece who lived at Canindé.

"But I've suruked her!" cried Antonio-hu in surprise.

"Pay me an arrow at once!" Chico said sternly.

"No, no, that was a long time ago!" True, he'd tried to make love to her again later, but she was pregnant and she wouldn't let him. 'Go away,' she said, 'don't you see my belly?'

"It's bad to suruk a woman when she's got a belly on her," said Antonio-hu, "if the child isn't yours. You go hunting in the woods and you don't see anything, you don't kill anything, you never kill anything again until the game speaks to you. But if it's your wife, your own child, then it's all right. Suruk another man's wife when she's with child and you see nothing but snakes in the woods, you get bitten by snakes."

"What happens if a woman can't make children?" I asked.

"I want to know, tell me! Before my wife died, I suruked her and made one child, my daughter. I suruked her again—nothing! Suruk suruk, nothing! Suruk suruk suruk, nothing happened." After that he made love to another woman whose husband wasn't living with her, and he had three children, two daughters and a son, but they died, they all died. Then his wife died too and left him with his daughter Ari, who he brought up all by himself.

"I couldn't make any more children in her," he said. "What's the remedy?"

I made quite sure that he didn't know, and then

hazarded the remark that in the old days men beat their barren wives with a snake, though what kind I wasn't sure.

"Was it spotted?" he asked, "or with stripes? . . . Stripes! black, and red, and yellow stripes, was it a mboi capitã? A long time ago yandé ramuĩ told me that when a man walks in the woods and sees mboi capitã (the coral snake), then he'll have lots of children. When you see one, don't kill it! It brings children into the world. Was it mboi capitã yandé ramuĩ beat his wife with?"

"Isn't mboi capitã poisonous?" I asked.

"No, you pick it up, it's all right." (Coral snakes can often be picked up without them biting you, though their venom is deadly.) Anyhow, Antonio-hu was going to beat Pinduaro with a mboi capitã when she grew up, if she didn't bear him children. "Mind you don't beat her with a jararaca, so that it bites her and she dies," said Chico. Antonio-hu nearly broke the hammock cord, laughing.

For Pinduaro, the nine-year-old daughter of Wirakangupik, was being brought up by Antonio-hu to be his wife. He was very pleased with the arrangement, the only drawback being that she was still much too young to sleep with; and it was no doubt partly because of this, because he felt lonely, that he had become so attached to Chico and me. His attitude to Pinduaro was that of a benevolent, if sometimes rather apprehensive uncle, for she was already complaining about his unfaithfulness. She had started that morning, when he told us what he had been dreaming about the night before: he had dreamt that he was courting a woman called Iro, who lived on the river Turi, some distance away. Iro had said, 'Let's go into the jungle to make love!' But Antonio-hu wouldn't, because in his dream he had a grown-up wife at home.

"Look at Pinduaro," said Antonio-hu, "surely she'll beat me when she grows up, already she scolds me because I dream of women!"

Pinduaro was in the flour house with Ari, baking flour. We wandered over to watch them. The soft, half-decayed mash of manioc tuber, squeezed in the woven tube or *tipiti* to get the poisonous juice out, was sieved to remove large fibres and put on top of the huge clay plate, under which a fire roared. Pinduaro was raking this with a flat rake, turning the flour over to stop it burning. When she saw Antonio-hu she began to scold him again in her squeaky voice, this time for another reason: the day before Saracaca had been making flour and he had stayed a long time raking the flour and talking to her, and had later given her a string of black beads. "You did, you did!" Pinduaro cried with the utmost self-possession, when Antonio-hu looked away and mumbled that he hadn't.

"You did," she repeated, "you'll turn into a cururu," and happily, energetically went on raking the flour. It was a serious charge, for a man only gives beads to a woman if he has slept with her (unless she is his wife), and Pinduaro was right in saying that Antonio-hu would become a cururu toad after his death for having committed adultery. Antonio-hu turned to me and began a lengthy explanation, but Pinduaro took no notice at all, just pushing him out of the way to place another log on the fire. "Cururu, cururu!" she repeated gaily. Her last shot at Antonio-hu was to say that Saracaca had a yahi-rata ramuĩ of a carapua, that is a grandfather of a star of a vagina. It was, I think, her own invention.

Pinduaro continued her mockery of grown-ups that evening, when she and Ari sat on the ground near Chico's hammock, calling out derisively, "Saracaca! Saracaca!" and twitting him over his attachment to her. Ari was particularly scathing, for she wanted a lover of her own, and thought that Saracaca had more than enough; so she told Chico to show some spunk, and make love to her instantly. Then the girls sang little songs, making up the words as they went along, all about Chico and how he only liked married women, how he wouldn't look at girls, wouldn't even flirt with them . . . Chico turned rather pink, but he enjoyed himself.

We spent three or four days in this way, looking at the rain fall, drinking shibé after shibé, and talking quietly together about nothing in particular. In the evenings Chico and Alishandre got together and sang shaman songs, the entire village acting as chorus. They sang better here than at Pari's, and the evenings were a success, for odd things happened—a young boy saw a sweet-smelling ghost, for example, and Antonio-hu saw a white spirit-snake with spots down its back. We would sing through the night, and then sleep till midday; then I would go out with Antonio-hu to shoot a howler monkey, perhaps, or Alishandre would go out by himself to hunt paca, and we would spend the rest of the day eating and talking.

On the last evening, we danced. It was a quite informal affair—no cauin, no feathers, no paint—for the dances themselves, though often performed at ceremonies, are nowadays usually done for amusement. A large fire was built in the open space between the huts, lighting everything up in its flickering, ruddy glare; then the girls, with Saracaca leading, got into a line and held hands. After a few giggles and nudges, Saracaca began to sing the duck song, which the others took up hesitantly, and the dance began. They bent their backs slightly, so that their hair fell forward and screened their faces, and paced forward; then they stopped, bowed up and down, retreated, and paced forward once more. After a little they gained confidence and began to sing the duck song as it should be sung, in high, full-blooded voices which from the repeated lower, breathy notes broke out into melodious whoops that were still part of the tune. The dance became less formal: Saracaca, leading, held her free arm above her head, the fingers pointing down—the duck's beak which, as she bowed and stepped back and forth, dipped sweetly to the ground, investigating a water-pot, a clump of weeds, the fire; the line of girls curving sinuously after her as she moved through the village. Then the boys had their turn, with Tero (who was then paying us a visit) in the lead. They moved through the shadows in a magnificent crouching lope, singing the yapu song: a song like the duck song, but with whoops louder and more resonant. The woods rang with echoes: the dust, kicked up from the ground and lit by the fire, swirled low in reddish clouds. Then, blessed joke, after an evening of summery airs and transparent sidelong light, the rain came down; it laid the dust flat, put out the fire, and sent the dancers running for shelter. "The rain falls," said Antonio-hu, peering out into the night.

The next day we were all rather silent. The day wore on slowly; I gathered my things together, untied my hammock, put it and my red flannelette blanket at the bottom of my rucksack, laid my typewriter on top and packed it about with a spare shirt and a pair of trousers; put my camera (broken, alas) in one of the pockets, and pulled on my gym-shoes.

"Chico," I called out, "I'm going."

Chico was sitting in the flour house, talking to Saracaca and Alishandre. He came over, looking sad.

"Have you any beads left anywhere?" he asked anxiously. I still had a handful; I gave him half, and the rest I doled out to Ari, and Pinduaro, and the rest of the children. A few empty tins that I had I gave to the men; Alishandre gave me a quantity of tawari bark for cigars, and some feathers. Saracaca looked at Chico as he rolled his hammock up, with tears in her eyes. "Don't you want a shibé?" she asked in a small voice.

"Don't you cry now! Yes, I want a shibé." So she went and made a large one, for Chico, Tero, me and

Antonio-hu, who was coming with us: and at about half-past four we set off down the path to Pari's.

It was the right time to travel: the rain had stopped at noon, and the sun had been out long enough to dry the jungle without taking away its freshness. The sunlight came slanting through the leaves in small patches, lighting up the small feathery ferns that had sprung up in the last month, the red flowers of the guaruma, the long pink tubes of the bromelias, and the dead leaves on the jungle floor; shone over the bacuri fruit lying rotten on the ground, on old tree-trunks covered in red and brown toad-stools, and through enormous spiders' webs. There were several bands of monkeys chittering in the branches, and many parrots and smaller birds volubly exclaiming as they ate fruit. I felt as though I was going on a holiday.

We stayed the night in Pari's hut, and ate the armadillo he had shot the previous day—very tender meat—and some bananas which he dug up from beneath the fire, where they had been ripening. Next morning we had a quick breakfast of my coffee and Pari's manioc flour, shouldered our loads and were on our way again just as the sun rose.

It had been raining in earnest the last month, but I hadn't realised quite how much. We saw, however, soon after we crossed the roça belonging to Canindé: though three-quarters of a mile distant from the river, the jungle there was flooded a foot deep. It was an extraordinary sight, to see this immense lake spreading under the trees; in it, the patchy reflections of the sky made the trees stand out so much that one could distinguish things twice as far away as usual. There was a curious silence, too, as though the jungle was waiting for something; a silence which the noise of an axe in the distance only made more obvious.

We waded along the path till we came to a canoe some-one working in the roça had tied up there. We piled our gear fore and aft, out of the bilge, and scrambled aboard—it was the largest dugout the post owned, and held us all easily. Chico and Pari, who had come too, took a paddle each, and moved the canoe on gently through the trees. Soon we were hilariously shouting, like a Bank Holiday party afloat on the Serpentine, making so much noise that I never heard Tero's yell of agony as a tocandira ant bit him: he'd put a hand right on to it when staving off the canoe from a branch. Our shouts were heard by the inhabitants of Canindé, on the other side of the river; they groaned at the prospect of feeding what sounded like two score Indians.

CHAPTER TWENTY-ONE

Indians and Brazilians

I WAS glad to get back to Canindé. The well-built house, the battery wireless João Carvalho switched on during the evenings, the meals served on china plates at a table covered by a cloth: these familiar comforts at once claimed me, and made the four months I had spent in the jungle curiously unreal. It was a joy to eat meat properly cooked once more, to taste fish, to have fried potatoes and salads and corn on the cob; there was even milk to be had, from the zebu cattle that wandered about round the post. Best of all was to be able to read in peace. I had managed to get through *Don Quixote* and *Moby Dick* while I was in the jungle, but I had read guiltily and without real enjoyment: I felt that to isolate myself with a book was somehow to cast a slight upon the Indians and their friendship for me. At Canindé, however, it was as if I was once more at home, and I read voraciously. I needed the change.

It was also strange to sleep in a room once more. Chico preferred to sleep on the verandah, where he could see the moon glinting on the river—it reminded him of his home at Jararaca, which we were soon to visit—but Antonio-hu slung his hammock inside, next to mine. It was not just because he liked being with me: he was beginning to feel out of things. We had, in fact, changed positions, and it was now his turn to endure the slight strain of living by an unaccustomed set of conventions: he found it especially irksome that conversations were now carried out in Portuguese and not in Urubu.

He had, however, plenty of company, for a number of Indians came to see João. Most of them brought loads of resin, or dried deer and peccary hides, which João would sell for them downriver; in exchange he would give them axes and machetes, tobacco and cloth. Among these Indians was a chief called Shapi, who arrived with his wife, his three sons, and several other men from his village. They had come empty-handed, but were all expecting to be given presents—indeed, they were loud-mouthed in their demands for machetes. "I'll bring you a lot of resin," Shapi said, "but now I want a machete." João, slightly mistrustful, gave him and the others some tobacco and a machete each; they went off and returned the next evening with one small load of resin that was nearly worthless. "Resin," said Shapi expansively.

I laughed, and in revenge João turned Shapi on to me. "Ask him for some beads," he told Shapi: and Shapi, an inveterate beggar, lost no time in asking. I refused. "My beads are finished," I said.

"But my wife wants some!"

"No more left!"

"Blue ones! You've got blue beads?"

"All gone!"

"My wife wants some, she longs for some."

"I haven't got any."

"No more? Blue ones! I want some too."

"No more!"

"Yes! In there!" (pointing to my suitcase).

"None left."

"In there, in there! Tero says you brought a great heap."

"What a lie, they're all finished."

"My wife wants some, look, my wife. She wants blue beads, she hasn't any at all."

"I haven't got any at all."

"All finished?"

"Finished."

"Finished?"

"Really finished."

"Finished," Shapi said wistfully.

His sons had their own kinds of beads—the youngest son, five or six years old, who had slant eyes and was very pretty, had the dried windpipe of a curassow bird round his neck, the eldest had curassow bones strung as a necklace: to make them into good hunters of curassow when they grew up. And one of the young men who came with Shapi had a necklace of jiboia snake rib-bones, to charm the game animals towards him. He also had the dried head of a young jiboia, which he sold to João for a length of cloth—an exorbitant price, João said, but his brother had asked him for a jiboia head because it's so good for hunting deer with.

Brazil owes a lot to the Indian, no matter how much he is derided. Indeed, were it not for the Indians' knowledge of the jungle, the Portuguese would have found it almost impossible to settle in Brazil. Cereal crops, which do not grow well in the tropics, had to be abandoned: instead the first settlers began to cultivate manioc, preparing it in the complicated and somewhat unlikely way the Indians had discovered. Nowadays, the Brazilian peasant has taken over many other Indian customs. He sleeps in a hammock; if poor, he uses a bow and arrow instead of a musket: he uses Indian game- and fish-traps; and he calls the animals, the birds and the plants by their Tupi names. With this knowledge of natural history goes a considerable amount of supernatural lore. The Brazilian peasant believes firmly in Curupir, the jungle spirit, in the concept of panem or magical bad luck, in the efficacy of tobacco smoke as a fumigant, and in shamanism; herbal remedies are nearly all based on Indian knowledge, and the best game charms, of course, are those made by the Indians.

One of these charms is even known in the cities. It is a bird called wira-puru, the jumping bird (*Pachysilvia* sp.):

the king of the birds, the Indians say, for wherever it goes there all the other small birds of the jungle follow it. It is therefore full of magic; and the Indians use it as a game charm if, when they kill it, it falls belly up on the ground. Belly down, it is used as a love charm, to be hidden in the hand, the beak suddenly bared and pointed at a woman when she's not looking. These birds are in great demand in Viseu and Belém: shopkeepers put them under their door-sills to ensure a flow of customers, business men keep them in their desks or their safes, lovers keep them in their pockets. There is such a demand for them that one Indian I knew went to Belém with twenty small dried birds of many different species, and managed to palm them all off as wira-puru; he came back equipped with new clothes, machetes, tobacco, and other fine goods.

But the things which Indians and Brazilians have in common mask real differences, both small and large. The small differences are absurdly trivial—such things as eating with the fingers instead of with a spoon, wearing shoes or going barefoot, saying 'give' and not 'please'—trivial, but still important, because there are so many of them. The large differences, however, are not so easily seen or understood: they are the particular values that every man takes for granted, the motives that give his life its meaning. In a word, it is a difference between religions; and it it is this which makes understanding between different peoples so difficult, for religion is something that can only be understood properly when it is lived.

I saw an amusing example of this difficulty at Canindé, when a child was christened. The father was not altogether happy about the situation, for the infant was not well—the mother had no milk left—he was sad, too, that he could not have the child baptised properly, by a priest, but that would have meant a journey downriver to Viseu, which he could not afford. He asked João to officiate instead.

The parents of the child had swept their hut, taken the hammocks down, and borrowed some benches which they put round the walls. Everyone who lived at Canindé came to see the ceremony: there was a dreadful crush, and João had a hard time getting inside the hut at all. At last he was ready to start. "Where's the water?" he asked. Cold water was brought. "No, make it warmer—do you want to freeze the poor thing?" Five minutes later a panful of boiling water was produced, brimful. "Oh, women, not all that. What's wrong with you all? Now the salt. Just a little." They brought him a plateful, from which he took a pinch to mix with the baptismal water, and another to pop into the baby's mouth later. João began the rite in a very low voice, the salt went into the baby's mouth and made it squall so much that Leudi, João's wife, fished it out again with a finger. João went on muttering rapidly, and at last came to an end; the child was given back to its mother, and the guests trooped off to the house where Leudi had cakes and coffee waiting.

Antonio-hu watched all this with bewilderment. What kind of ceremony was it where no one dressed up in feathers, no one got drunk, no one really heard what the baby's name was or repeated it after the godfather, to make sure he knew it: where no one had danced or sung? "Is that all?" he asked me. But Leudi was pleased—"Just like a priest!" she complimented her husband.

Antonio-hu could see the point of giving a child a name, and doubtless if it had been explained to him he would have understood the meaning of the baptismal water and the purifying, truth-giving salt: but he could hardly have understood the Christian reality behind those symbols, since there was nothing in his own mythology corresponding to it. Missionaries have always found it difficult to make primitive peoples understand what is being referred to under the names of God, Christ, and of various saints: and they have often found it useful to take over some figure from the aboriginal mythology and elevate him to the status of God—the others, inevitably, becoming devils. In Brazil, missionaries have always referred to God by the name of Tupan, the thunder. Thunder-gods have a way of becoming Absolutes, no doubt because they make so much noise, and blast people with their bolts:* and Tupan was in any case a better choice than Maír, so plainly a superman, for Tupan has the theological advantage of being disembodied and of living in the sky.

But names are powerful things, especially when they refer to mythological forces: Tupan, though an effort has been made to turn him into a Christian, has sometimes proved quite contrary, and turned Christ into an Indian instead. This has happened among the Urubus. An English mission worked near Jararaca a number of years ago—it has since been disbanded—and, it seems, rivalled the Indian post in popularity, for Indians were continually going there either for gifts, or for medicines, or just for the fun of it. Though it made no converts among them (as far as I know), it did much to help the Indians and from its teaching there arose a completely new myth, which Anakãpuku told me. I had asked him who it was that had made dogs in the beginning of the world: to my surprise, this is what he said.

Tupan came down to earth one day, and suruked a married woman. She didn't say anything to her husband about this, but some months later he couldn't help noticing that she was pregnant. 'That's not my child!' he said angrily. At last the child was born, and one morning the woman left it in her hut while she went down to the river to get water. The husband took his chance: he beat the child, stabbed it with his knife, and hid it under a stone.

* It is interesting that throughout the world stone axes are thought to be thunderbolts. The Urubus have the same belief, even though some of them know that in the old days their ancestors made these 'thunderbolts' and used them to cut down trees.

The woman came back from the river—where was her child? 'Child, child, where are you?' she cried, and the child answered from under the stone, 'Mother, here I am.' There was her child, bruised and bleeding. But Maír had been watching, and took pity on the child: he came, and puffed tobacco smoke over it, so that it recovered. 'Mother,' the child said then, 'if I stay there'll be other people coming to kill me: I'm going off.' So he got on to a kind of chair Maír had provided, a chair like an aeroplane, and was carried off into the sky, where he now lives. And from the blood he shed were born dogs, and chickens, and ducks: all animals that have been introduced from Europe.

Chico and I saw the site where the mission had stood when we made our way upriver to Jararaca. The jungle was already growing up thick and tall upon it; I should have missed it if Chico hadn't pointed it out to me. Jararaca, however, was in better shape, even though the Indian post that had been started there for the Tembés was no longer functioning, for a number of Tembés still lived there. Their village was built in a most attractive place, on a bluff overlooking a loop of the river: there was a row of palm trees on the bluff, and behind them a great thicket of guava bushes, heavy with fruit; in the distance one could see the dark green tops of a grove of mango trees.

Chico was glad to be back with his family once more; he had been getting homesick. He took me round to see another small village of Tembés, an hour's march away, and the neighbouring Urubu village where I met Kuashipuru, who told me the extraordinary rite the tushau has to undergo after he has ritually killed an enemy prisoner.

I stayed some six weeks at Jararaca, enjoying myself a great deal. It was the beginning of summer: the rains were ending, and the days were becoming hot and golden once again. Chico enjoyed himself too: he would sit on the bluff, singing to himself and enjoying the wind that now blew freshly all day. 'O verãozão,' he would exclaim, 'o *summer*.' His favourite song at the time was a local samba that started like this—

O menino de hoje
E uma ponta dificil,
Quando cresce um bocadinho
Quer namorar . . .

The boy of today
Is a difficult case,
When he grows just a little
He wants to go courting . . .

Chico had grown up quite a lot. After a week at Jararaca his Tembé blood asserted itself, and he decided to give up the idea of living with Saracaca—"Perhaps she won't come here if I ask her to; besides," he said, in a final and damning judgment, "she can't sing." He began to flirt assiduously with a number of young women, and before a month was out had been caught in an engagement by a Tembé girl called Jelsa who, besides being extremely handsome, did know how to sing: her father, like nearly all Tembé men, was something of a shaman. So Chico began to make great plans for the future, and drew up a list of things he wanted to buy in Belém—he was coming with me to see the city and to lay out his salary to the best advantage. (A lot of his money went on perfume and hair-oil, in elegant bottles he could not resist; but, he told me somewhat shamefaced, he could always sell them at three times their original price anywhere on the Gurupí.)

Conveniently for us, Feliciano was just about to make another journey downstream to Viseu, and he agreed to give us a lift. He came floating down the river one afternoon in Mundik Tavares' other boat, a large launch, with a couple of men paddling, for he had run out of petrol.

With Feliciano came dona Benedita, garrulous as ever, with her son Dorico. She had got bored with life at Mundik's lumber camp, and wanted a change.

Kuashi-puru and a couple of other men from his village had come to see me off: they and the Tembés helped to carry my luggage down to the river and load it aboard the launch.

"Are those the caboclos you stayed with?" dona Benedita asked, using the slightly derogatory name by which Brazilians refer to the Indians. Standing on the deck of the launch as we pushed off and floated downstream, I looked at them affectionately and waved my hand. Alas, compared to Dorico in his newly-ironed shirt and trousers, they did not look at all respectable: but nor, I comforted myself, did I.

Glossary

All Urubu words are accented on the last syllable. The pronunciation of the vowels is as follows:

a as the u in *sun*
ã nasalised; as in *sung*
e as in *pen*
é as in the French *thé*
i as the ee in *reed*
ĩ nasalised; somewhat as in *sing*
o as in *for*
õ nasalised; as in *song*
u as the o in *do*

U	Urubu
B	Brazilian
T	Tembé
Tu	Tupi

Aé. A cannibal spirit. (U)
Anawira. A subsidiary culture-hero; the anawira tree. (U)
ang. Soul, shadow. (U)
Anyang. The spirit(s) of the dead. (U)
Ara-yar. Macaw-owner. One of Maír's epithets. (U)

caá-por. An Indian; the name by which the Urubus call themselves. From caá, jungle, and -por, inhabitant. (U)
caboclo. Originally, an Indian-negro half-caste, and thus a peasant; also used to refer to Indians. (B)
capitã. A chief. Derived from capitão. (U)
capitão. Chief, captain. (B)
capoeira. A deserted plantation. (B)
carapua. Vagina; from cara, potato, and puá, round. (U)

cararan. Penis; from cara, potato, and ran, similar to. (U)
caruwa. A magic spell or spirit token. (U)
cauin. Fermented beer made from manioc or fruit; from kaú, drunk, crazy, and i, water. (U)
Curupir. The jungle spirit; from curu, rough, and pir, skin. (U)

hantan. Hard, strong. (U)
hu-hu! Intimidatory shout of an angry Indian. (U)

ita kitik. Clashing rocks or Symplegades; from ita, stone, and kitik, to rub. (U)
ita kuí. The same; from ita, stone, and kuí, to grind. (U)
iwi pita. The end of the world, where Maír lives; from iwi, earth, and pita, to remain. (U)
Iwitu ràmuĩ. Grandfather Wind, a mythological figure. (U)

katu. Good. (U)
kaú. To be drunk, mad, crazy. (U)
kwa. Exclamation. (T)

mai, *mai-té.* Exclamation of wonder; from mai, a thing, and té, true, indeed. (U)
Maír. The culture-hero. (U)
Maír mimi. Maír's son. (U)
Mamai-hu. Big Mother, the Queen. See Papai-hu. (U)
maraca. A gourd rattle used by shamans. (U)
membek. Soft, weak. (U)
Mikur. The opossum. The shaman's familiar spirit; Maír's mortal counterpart. (U)
Mikur mimi. Mikur's son. (U)
mimi. Child, semen. (U)

nde, *ne.* You. (U)
ne'eng hanten. Hard speech: the formal language used by men when visiting each other. From ne'eng, speech, and hantan, hard, difficult. (U)

pai-té. Exclamation of wonder; see mai. (T)
panem. To fail, to be unlucky. (U)
Papai-hu. Big Father: the spiritual head of any civilised nation, such as the President of Brazil, or the King of England. From papai, father, and -hu, big. (U)
patuc. To stamp: a display of stamping and shouting put on when an Indian is angry. (U)
payé. A shaman. (U)
pera. A knapsack of plaited palm-leaves used for carrying game, fruit, etc. (T)
porang. Beautiful, virtuous. (U)

raír. Child, semen. (U)
rankuaĩ. Penis. (U)
Rankuaĩ-ang. A ghost or shadow penis. (U)
roça. A plantation. (B)

saé. Man, fellow. (U)
shibé. A drink of manioc flour steeped in water. (B)
suruk. To make love to a woman. Literally: to pierce. (U)

taboca. Bamboo; also an arrow with a leaf-shaped bamboo head. (U)
takwara. An iron-headed arrow. (U)
tamuĩ. Grandfather. (U)
tenetehar. A were-animal. Literally, a relation. (U)
Timakanã. A jungle spirit. From tima, leg, akang, bone, and ang, ghost. (U)
tobajara. Brother-in-law; the name of an enemy tribe. (Tu)

Urubu. From uruhu, a vulture. (U)

yandé ari. Our grandmother. (U)
yandé ramuĩ. Our grandfather, or forefather. (U)
ya té, ya té gui. Affirmatory ejaculation; literally, a true word indeed. (U)

KINSHIP CHART

The chart on the facing page shows how the members of Pari's village and Antonio-hu's village are related. Names in parentheses are relatives in other villages.

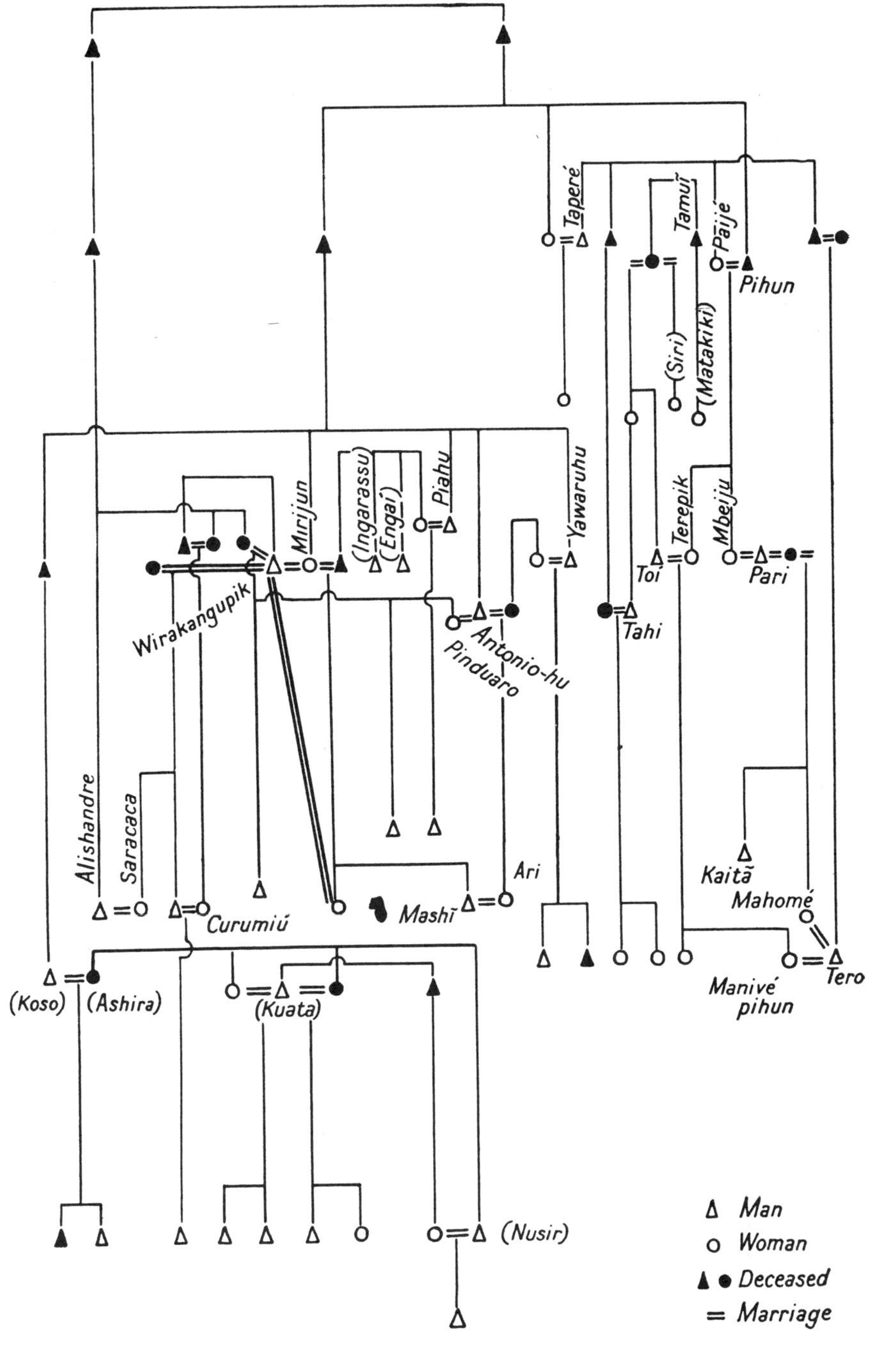

Taperé
Tamuĩ
Pãije
Pihun
(Siri)
(Matakiki)
(Ingarassu)
(Engai)
Piahu
Mirijun
Yawaruhu
Terepik
Mbeiju
Toi
Pari
Wirakangupik
Antonio-hu
Pinduaro
Tahi
Alishandre
Saracaca
Ari
Kaitã
Mahomé
Mashĩ
Curumiú
Manivé
pihun
Tero
(Koso)
(Ashira)
(Kuata)
(Nusir)
Man
Woman
Deceased
= Marriage